57 **Topics in Current Chemistry**

Fortschritte der chemischen Forschung

Cyclic Compounds

Springer-Verlag Berlin Heidelberg GmbH

This series presents critical reviews of the present position and future trends in modern chemical research. It is addressed to all research and industrial chemists who wish to keep abreast of advances in their subject.

As a rule, contributions are specially commissioned. The editors and publishers will, however, always be pleased to receive suggestions and supplementary information. Papers are accepted for "Topics in Current Chemistry" in either German or English.

ISBN 978-3-662-15512-7   ISBN 978-3-540-37564-7 (eBook)
DOI 10.1007/978-3-540-37564-7

Library of Congress Cataloging in Publication Data. Main entry under title: Cyclic compounds. (Topics in current chemistry; 57). Bibliography: p. Includes index. CONTENTS: Eicher, T. and Weber, J. L. Structure and reactivity of cyclopropenones and triafulvenes. – Sargent, M. V. and Cresp, T. M. The higher annulenones. 1. Cyclic compounds – Addresses, essays, lectures. I. Eicher, Theophil, 1932-   II. Series. QD1.F58   vol. 57 [QD331]   540'.8s [547'.5]   75-11665   ISBN 978-3-662-15512-7

© Springer-Verlag Berlin Heidelberg 1975
Originally published by Springer-Verlag Berlin Heidelberg New York in 1975
Softcover reprint of the hardcover 1st edition 1975

Typesetting and printing: Schwetzinger Verlagsdruckerei GmbH, Schwetzingen.

Contents

Structure and Reactivity of Cyclopropenones and Triafulvenes

Prof. Dr. Theophil Eicher and Dipl.-Chem. Josef L. Weber

Institut für Organische Chemie und Biochemie der Universität, D-2000 Hamburg 13, Germany

Contents

I. Introduction

The field of "microcyclic compounds"[1] has become highly attractive in the past few years through the intriguing, and in many cases unexpected, structural properties and chemical reactivity of these compounds. Among them, cyclopropenones *1* and their functional derivatives *2–4* as well as triafulvenes *5*[2, 3] have become of considerable interest in preparative and theoretical chemistry and have been documented in a number of reviews dealing with single members of this family of cross-conjugated cyclopropene derivatives or with partial aspects of their chemical behavior[4–12].

1: X = 0
2: X = S
3: X = N–R
4: X = N⟨R R' (⊕)

5

In this article it is intended to give a collective and comparative survey of cyclopropenone and triafulvene chemistry in the following areas:

1. synthetic approaches to cyclopropenones, their derivatives, and the various types of triafulvenes,
2. investigations dealing with their molecular and electronic structure as well as their spectroscopic and other physical properties,
3. reactivity and preparative potentiality classified according to aspects of reaction mechanism.

The literature up to the end of August 1974 has been covered.

II. Synthesis of Cyclopropenones and Triafulvenes

1. Cyclopropenones

The origin of cyclopropenone chemistry goes back to the successful preparation of stable derivatives of the cyclopropenium cation *6*[13], the first member of a series of "Hückel-aromatic" monocyclic carbo-cations possessing a delocalized system of $(4n + 2)$-π-electrons. This experimental confirmation of LCAO-MO theory stimulated efforts to prepare other species formally related to cyclopropenium cation by a simple resonance description of electron distribution, namely cyclopropenone *7* and methylene cyclopropene (triafulvene) *8*:

$(4n+2)\pi$
with $n=0$

X=0: 7
X=CH$_2$: 8

6   a   b

3

It was expected that participation of the dipolar forms b in the ground-state hybrid of 7/8 might introduce the special electronic stability of the delocalized 2π-configuration to compensate for the high strain energy estimated for these molecules, (e.g. 8: 58 kcal/mole[14]. From calculations according to the HMO model considerable delocalization energies were predicted for cyclopropenone ($DE = 1.36\,\beta$) and its phenyl-substituted analogues (phenyl cyclopropenone: $DE = 3.75\,\beta$, ΔDE to 7: $0.39\,\beta$; diphenyl cyclopropenone: $DE = 6.15\,\beta$, ΔDE to 7: $0.79\,\beta$[5]).

The isolation of cyclopropenones and their undoubtedly increased stability compared to the less-strained saturated cyclopropanones might well be attributed to the validity of the above symbolism of "aromatic" cyclopropenium contribution to the ground state of 7. It should nevertheless be clear, that the information available on the electronic structure of cyclopropenones demands certain refinements of this very useful qualitative concept.

a) Syntheses by Carbene Methods

The first synthesis of a cyclopropenone was reported in 1959 by Breslow[15], who achieved the preparation of diphenyl cyclopropenone (11) by reacting phenyl ketene dimethylacetal with benzal chloride/K-tert.-butoxide. The phenyl chloro carbene primarily generated adds to the electron-rich ketene acetal double bond to form the chlorocyclopropanone ketal 9, which undergoes β-elimination of HCl to diphenyl cyclopropenone ketal 10. Final hydrolysis yields 11 as a well-defined compound which is stable up to the melting point (120–121 °C).

$Ph=C_6H_5$

9   10   11

This method has also been used for the preparation of a series of aryl phenyl cyclopropenones[16].

Independently Volpin[17] synthesized diphenyl cyclopropenone from diphenyl-acetylene and dibromo carbene ($CHBr_3$/K-tert.-butoxide). This reaction principle of (2 + 1) cycloaddition of dihalocarbenes or appropriate carbene sources ("carbenoids") to acetylenic triple bonds followed by hydrolysis was developed to a general synthesis

$R-C{\equiv}C-R' + ICX_2$ 12   13

R, R'=aryl, alkyl, not H
X=F, Cl, Br

of disubstituted cyclopropenones *13* when applied to dialkyl- or diarylsubstituted acetylenes. The required dihalocarbenes can be generated from various combinations of haloform/K-tert.-butoxide or methyl trichloroacetate/NaOCH$_3$ in an inert solvent or — as recently found[18] — in a two-phase system by phase-transfer catalysis. The dihalocarbene method is convenient, but the yields of cyclopropenones are only moderate (15–25%).

The reactivity of dichloro carbene towards acetylenic bonds was systematically investigated by Dehmlow[19, 20] with respect to substitution of the acetylene, especially those containing additional C-C multiple bonds. It was shown that with aryl alkyl acetylenes, *e.g.* 1-phenyl-butyne-1, often the "normal" cyclopropenone formation occurs only to a minor extent (to yield, *e.g. 14*), whilst the main reaction consists of an insertion of a second carbene moiety into the original acetylene-alkyl bond (giving, *e.g. 15*):

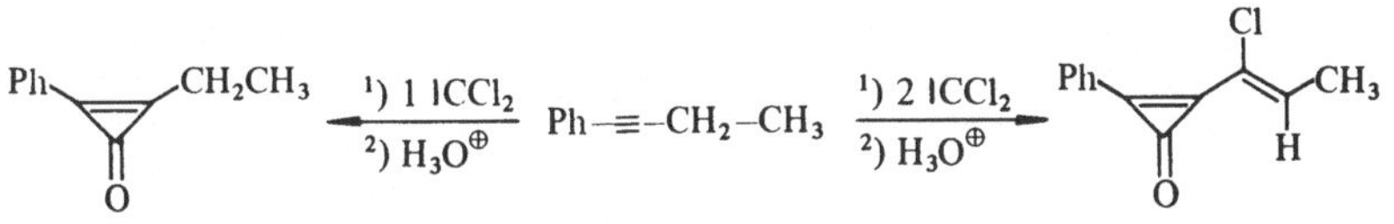

Furthermore, the addition of dichlorocarbene to ene-ynes proved to be remarkably sensitive to substituent effects. Trans-1,4-diphenyl butenyne gave only the cyclopropenone *17* via hydrolysis of dichlorocyclopropene *16*, however, 2-methyl-pentene-1-yne-3 favored the formation of the dichlorocyclopropane *18* with only traces of products resulting from addition to the triple bond:

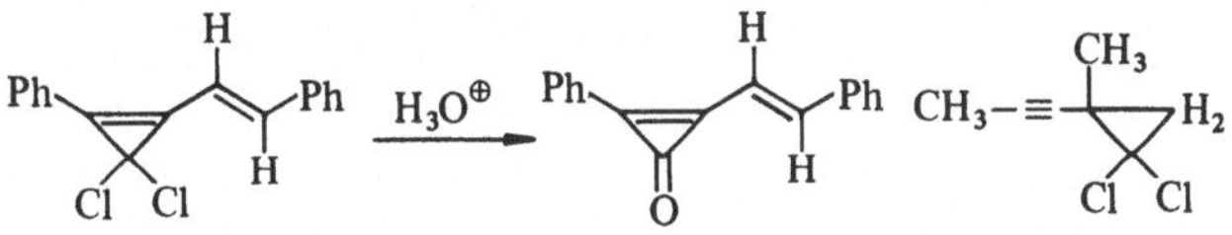

The dihalocarbene method was expanded in scope and improved in yield by the introduction of carbenoids of several types. Thus, (dichloro-bromomethyl) phenyl mercury reacted with diaryl acetylenes giving diaryl cyclopropenones in high yields (60–80%)[21] after hydrolytic work-up. The corresponding tribromomethyl mercury compound also reacted with aryl alkyl acetylenes[22].

Trichloromethyl lithium (generated from BrCCl$_3$ and CH$_3$Li at −100 °C) adds to dialkyl acetylenes and to monoalkyl acetylenes[23], thus monoalkyl cyclopropenones became accessible which could not be obtained from terminal acetylenes by reaction with the above carbene sources. The 3,3-dihalogeno-$\Delta^{1,2}$-cyclopropenes formed as primary products in the dihalocarbene reactions are usually not isolated, but are hydrolyzed directly to cyclopropenones.

Finally, the addition of difluorocarbene (generated from F$_2$ClCCO$_2$Na) to steroidal acetylenes has been achieved[24−26] giving rise to exotic cyclopropenones bearing steroid systems as substituents, *e.g. 19/20*.

19 (R=CH₃, H)   20

The preparation of *unsubstituted cyclopropenone* (7)[27, 28] was achieved by reducing the readily available tetrachlorocyclopropene (21)[29] with tris(n-butyl)-stannane; this produced 3,3-dichloro-Δ^{1,2}-cyclopropene (22), whose carefully controlled hydrolysis gave rise to a 41—46% yield of pure cyclopropenone, which was stable under an inert gas atmosphere at temperatures below its melting point of −21 °C.

A further convenient preparation of 7 was found[30] to be the hydrolysis of cyclopropenone dimethyl ketal (23)[31].

21   22   7   23

24   25   26

Application of the usual hydrolysis procedures to tetrachlorocyclopropene does not lead to the formation of dichloro cyclopropenone (26). This unstable compound is obtained, however, by a special procedure from trichloro cyclopropenium tetra-chloroaluminate (24) via the AlCl₃-adduct 25[32].

b) Trichlorocyclopropenium Cation Method

A very valuable method for the synthesis of arylsubstituted cyclopropenones was found by West[33–37]. In this method the trichloro cyclopropenium cation 24 or tetrachlorocyclopropene in the presence of Lewis acids like AlCl₃ or SbCl₅ is reacted with benzene or derivatives of benzene bearing alkyl, alkoxy, halogen, or hydroxyl substituents. Although the introduction of three aromatic residues is possible[36], the arylation of 24 can be conducted preferentially to the stage of the diarylsubstituted cation 28, whose hydrolysis gives rise to diaryl cyclopropenones 29. In some cases arylation can be stopped at the monoaryl stage 27, thus giving rise to the formation of aryl chloro cyclopropenone 31 and — through reaction of 27 with a second aromatic species — to the introduction of different aromatic residues.

Finally, the alkylation could be extended to vinyl halogen compounds[38] and in this way, bis(trichlorovinyl)cyclopropenone (*30*) was prepared from cation *24* and trichloroethylene. A recently reported variation of the West procedure[40] makes use of the difluoro chloro cyclopropenium cation *33*.

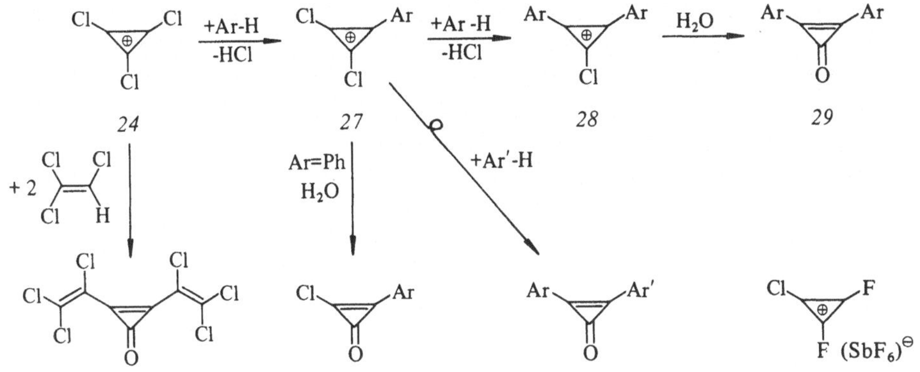

24 *27* *28* *29* *30* *31* *32* *33*

Monosubstituted benzene derivatives are generally attacked in the para-position by trichlorocyclopropenium cation; for toluene it was recently found[39] that ortho-attack also may occur as indicated by formation of *35* besides the "normal" product *34*:

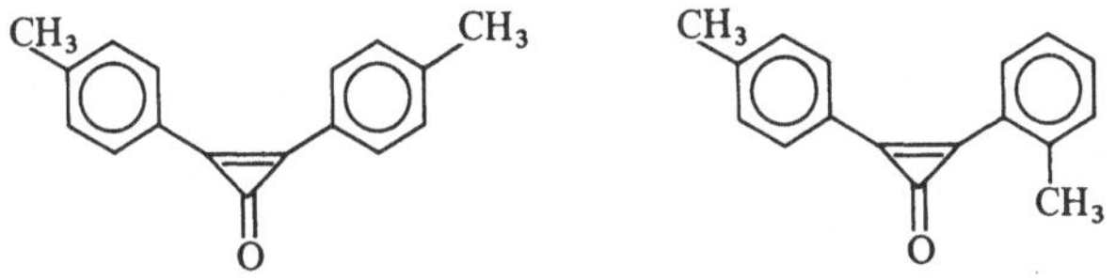

34 *35*

c) Dibromoketone Method

In 1963 Breslow published a synthesis of cyclopropenones, which approached formation of the three-ring without the use of divalent carbon species[41–45]. α, α′-Dibromo ketones *36*, in general readily available from ketones R–CH$_2$–CO–CH$_2$–R′, are de-

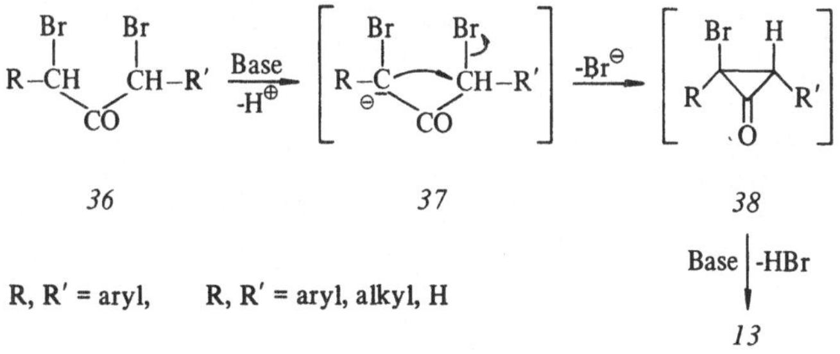

36 *37* *38*

R, R′ = aryl, R, R′ = aryl, alkyl, H

hydrohalogenated in a Favorski-type reaction by means of tertiary bases like triethyl-amine or diisopropylethylamine[46] as well as K-tert.-butoxide. In the primary 1,3-elimination process (via the anion *37*) formation of a bromocyclopropanone *38* is postulated, which undergoes β-elimination of HBr to the cyclopropenone *13*.

This procedure is the method of choice for the preparative chemist, since it not only provides the advantage of generally good yields (40–60%), but also can be carried out on a large scale. The modified Favorski reaction of dibromoketones offers a remarkably wide scope of application and has made possible the preparation of a large number of dialkyl-, diaryl-, and monoaryl cyclopropenones (see Table 1). Interestingly, the bicyclic cyclopropenones *39*[43] and *40*[47] have become accessible by use of the dibromoketone method.

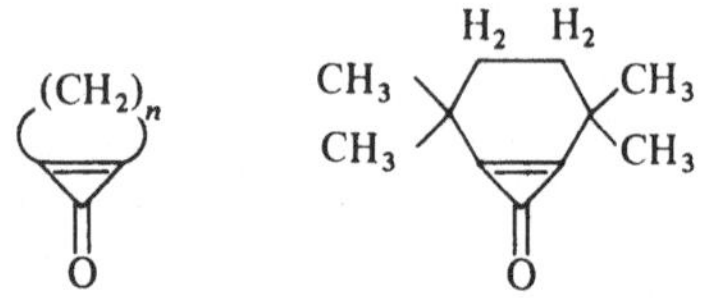

39 (n=5, 9)   *40*

Geminal α-dibromoketones *41* were also found[48] to undergo dehydrohalogenation to monoalkyl cyclopropenones *42*:

$$R-CH_2-CO-CHBr_2 \xrightarrow[-2HBr]{Base}$$

41   *42*   R=alkyl

d) Other Approaches to Cyclopropenones

Two other methods of cyclopropenone formation have become known; although interesting in principle they have not yet found general application.

Dehmlow[49] found that the photochemical extrusion of carbon monoxide from the cyclobutene dione system is possible as exemplified by the conversion of diethyl squarate (*43*) to diethoxy cyclopropenone (*44*):

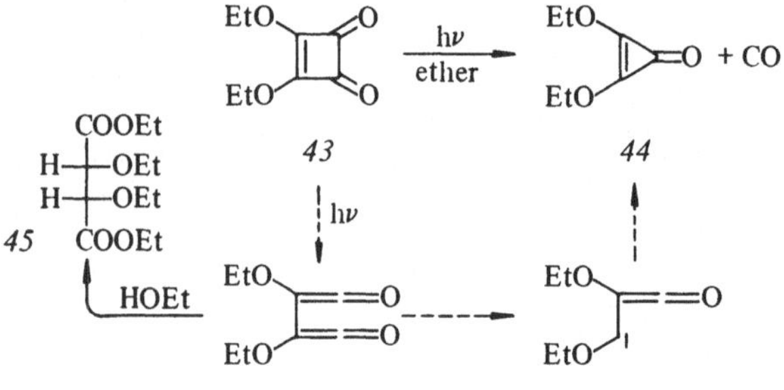

43   *44*

45

46   *47*

Cyclopropenone formation should involve the bisketene *46* and its decarbonylation to the "monoketene" *47* (a valence tautomer of cyclopropenone *44*), since photolysis in protic media like ethanol produces diethoxy diethyl tartrate (*45*) (meso and D,L). This method was also successful in the case of 1,2-diphenyl-3,3-dichloro-cyclobutene dione (*48*) giving rise to diphenyl cyclopropenone[49] (but as for *44* only a moderate yield was produced):

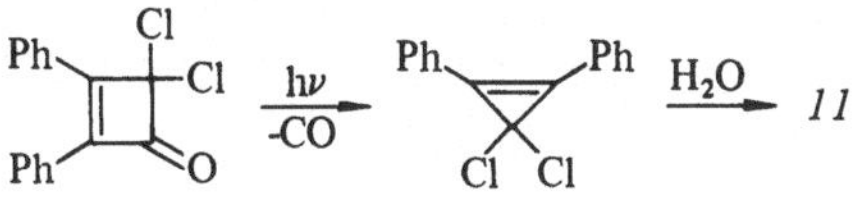

48

The formal relationship between cyclopropenone and an α,α'-biscarbene of a ketone ($R-\bar{C}-CO-\bar{C}-R'$) initiated investigations on photolytic and Ag-catalyzed decomposition of α,α'-bisdiazo dibenzyl ketone (*49*) (Trost[50]). Indeed, diphenylcyclopropenone was formed in addition to other products (*52* and tolane) derived from it; furthermore, products resulting from solvent insertion and Wolff rearrangement of the monocarbene *50* were isolated (*51*):

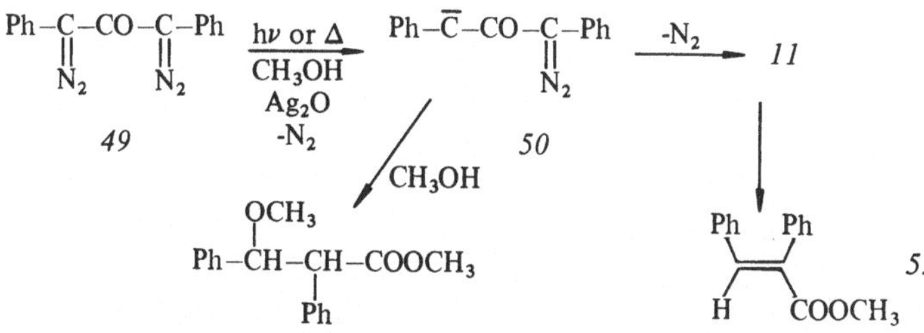

49 *50* *52*

51 (threo and erythro) + tolane

The theoretically interesting phenyl hydroxy cyclopropenone (*57*) was prepared by Farnum[51, 52] according to the general principle of cyclopropene ring closure developed by Closs[53] from *53* via the vinyl carbene *54* and phenyl trichloro cyclopropene (*55*).

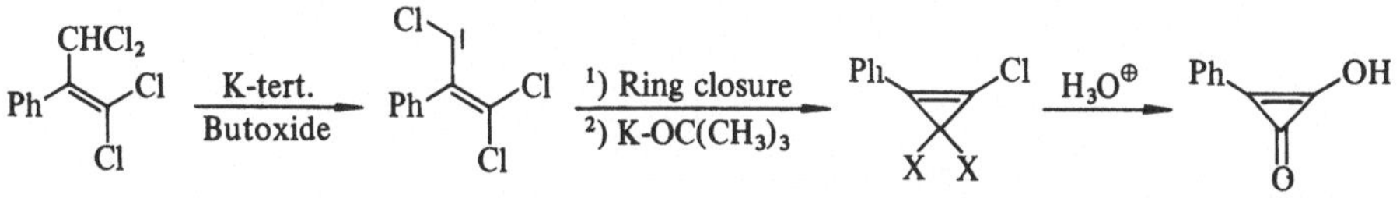

53 *54* *55* : X=Cl *57*

56 : X=OC(CH₃)₃

In an alternative synthesis[32, 33] *55* can be obtained [see (3)] from trichloro cyclopropenium cation and benzene. The yields of *57* (strong acid, $pK_a = 2.0 \pm 0.5$) are in the range of 20%.

Table 1. Preparation of some cyclopropenones and comparison of methods applied[1])

Cyclopropenone		Yield (%)	Method[2])	Refs.
H⎯▽⎯R (O)	R = CH_3	20	(1)	23)
	R = $n-C_3H_7$	17	(1)	23)
	R = $n-C_5H_{11}$	15	(3)	48)
(steroid, OAc/AcO)	R =	59	(1)	26)
(steroid, OAc/AcO)	R =	15	(1)	26)
H⎯▽⎯Ar (O)	Ar = C_6H_5	33	(3)	54)
	Ar = $p-CH_3O-C_6H_4$	14	(3)	55)
R⎯▽⎯R (O)	R = CH_3	19	(1)	23)
		12	(3)	56)
	R = $n-C_3H_7$	9	(1)	43)
		9	(3)	43)
	R = $n-C_4H_9$	12	(3)	43)
	R = $C(CH_3)_3$	36	(3)	44)
	R = cyclopropyl	6	(3)	57)
	R = $-CCl=CCl_2$	47	(2)	38)
	R = $-CH=C(CH_3)_2$	–	–	58)
	R = Cl	–	–	32)
	R = OC_2H_5	10	–	49)
	R = $-N(CH(CH_3)_2)_2$	80	(2)	64)
	R = $-S-C_6H_5$	4	(1)	62)
(X ring) (O)	X = $-(CH_2)_5-$	56	(3)	43)
	X = $-(CH_2)_9-$	8	(3)	43)
	X = $-\underset{CH_3}{\overset{CH_3}{C}}-CH_2-CH_2-\underset{CH_3}{\overset{CH_3}{C}}-$	41	(3)	47)
Ar⎯▽⎯Ar (O)	Ar = phenyl from phenylketene dimethyl acetal	80		42)
	diphenyl acetylene	24		42)
		63		21)
	α,α'-dibromo dibenzyl ketone	45		42)
		60		59)
	α,α'-dichloro dibenzyl ketone	12		42)

Table 1 (continued)

Cyclopropenone		Yield (%)	Method	Refs.
	Ar = p-tolyl	33	(2)	60)
	Ar = p-anisyl	73	(2)	60)
	Ar = p-tert.-butyl-phenyl	60	(2)	61)
	Ar = p-fluoro-phenyl	49	(2)	33)
	Ar = p-chloro-phenyl	53	(2)	60)
	Ar = 3,5-diisopropyl- -4-hydroxy-phenyl	82	(2)	35)
Ph⟍△R (C=O)	R = CH_3	70	(3)	45)
	R = C_2H_5	44	(3)	62)
	R = $C(CH_3)_3$	65	(3)	63)
	R = $CH_2-C_6H_5$	23	(3)	62)
	R = 1-naphthyl	54	(3)	63)
	R = trans-β-styryl	10	(1)	62)
	R = phenylethynyl	6	(1)	62)
	R = Cl	20	(2)	37)
	R = OH	36	(2)	37)
	R = $N(C_2H_5)_2$	–	(3)	43)

[1]) Further cyclopropenones of more complex substitution are compiled in Ref.[10]).
[2]) (1) = Dihalocarbene method.
(2) = Trichlorocyclopropenium cation method.
(3) = Dibromoketone method.

2. Triafulvenes

Methylene cyclopropene (8), the simplest triafulvene, is predicted to be of very low stability. From different MO calculations[5]) it has been estimated to possess only minor resonance stabilization ranging to ~ 1 β. Its high index of free valency[4]) at the exocyclic carbon atom causes an extreme tendency to polymerize, a process favored additionally by release of strain. Thus it is not surprising that only one attempt to prepare this elusive C_4H_4-hydrocarbon can be found in the literature. Photolysis and flash vacuum pyrolysis of cis-1-methylene-cyclopropene-2,3-dicarboxylic an-hydride (58), however, did not yield methylene cyclopropene, but only vinyl acety-lene as its (formal) product of isomerization in addition to small amounts of acety-lene and methyl acetylene[65]):

58

Stabilization of the methylene cyclopropene system has to be expected according to the introductory concept of triafulvene stabilization, if: (a) the exocyclic C-atom

11

is part of a system delocalizing the negative charge, or if (b) the three-membered ring is part of a system delocalizing the positive charge. Accordingly, four types of resonance-stabilized derivatives of methylene cyclopropene have been realized thus far:

a) triafulvenes *59* with the exocyclic C-atom bearing electron-withdrawing substituents like CN, COOR, COR etc.,

b) calicenes *60* which have the exocyclic C-atom incorporated into a cyclopentadienyl system,

c) quinocyclopropenes *61* in which the exocyclic C-atom is part of a quinomethane (X = O) or quinodimethane (X = CR$_2$) system,

d) cyclopropenium cyanines *62* which combine the three-ring with another donor group by an appropriate number of methine groups.

It should be pointed out that there are some methylene cyclopropene derivatives, whose stability is ascribed mainly to inductive effects brought about by strongly electron-withdrawing substituents. Thus, 1,2-bis(p-tolyl)-4,4-(bis-trifluoromethyl)-triafulvene (*63*) synthesized recently by Agranat[66] is a perfectly stable molecule with a dipole moment (7.42 D) comparable to that of 1,2-diphenyl-4,4-dicyano-triafulvene (*64*) of the resonance-stabilized type (1)[67] (7.9 D).

Likewise, triafulvenes *65/66* substituted only by fluorine are reported to be stable[68, 24].

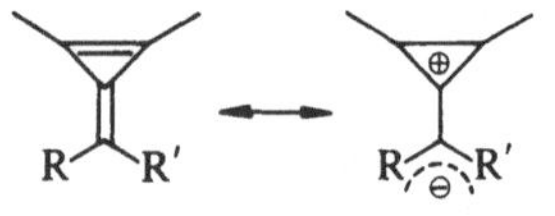

59 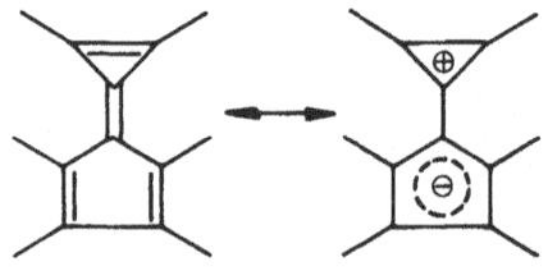   *60*

61a : X=O
61b : X=C(R)(R')

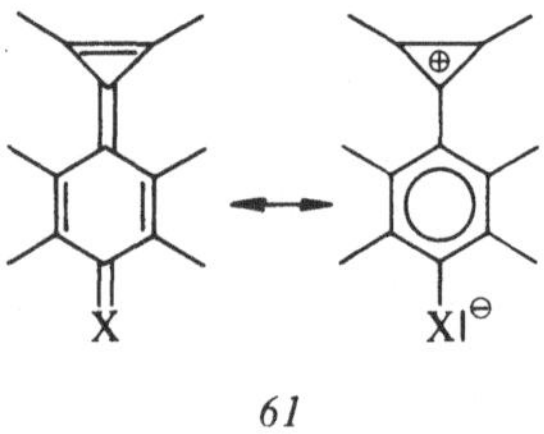

61 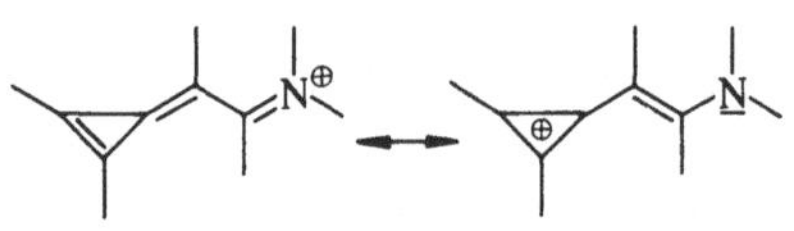   *62*

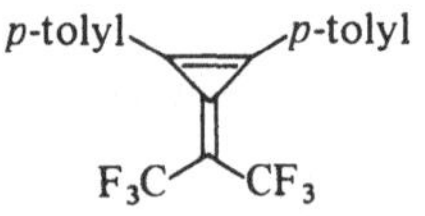

63 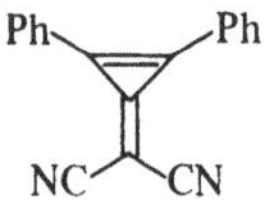   *64* 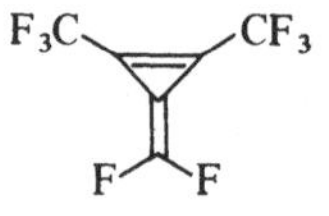   *65* 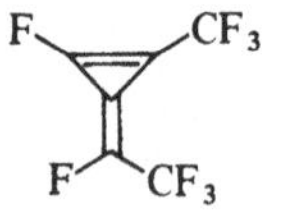   *66*

a) Methylene Cyclopropenes

One of the first syntheses of a triafulvene utilized the Wittig reaction, when diphenyl cyclopropenone was reacted with triphenyl carbomethoxymethylene phosphorane giving *69*[69]:

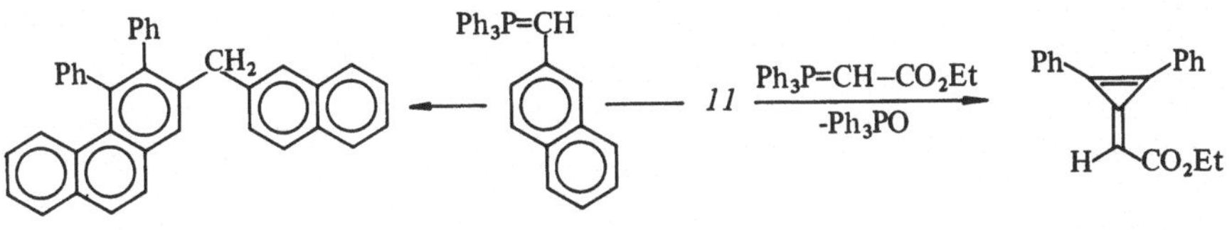

67   68   69

The inverse functionalization of the two components for a Wittig reaction has been described by Russian authors[70]; who combined the cyclopropenylide 70 with aldehydes to give the unstable methylene cyclopropenes 71 characterized by protonation as cyclopropenium salts 72:

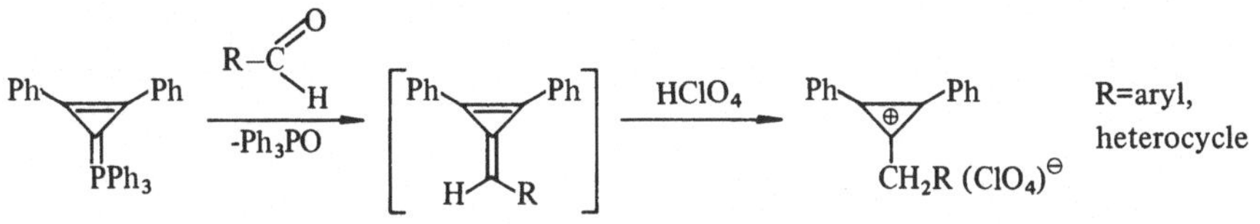

70   71   72

The Wittig reaction of diphenyl cyclopropenone with the phosphorane 68 failed to give the methylene cyclopropene; instead a hydrocarbon of probable structure 67 was obtained[71].

Triafulvene 69 was synthesized by an interesting route starting with diphenyl cyclopropenium cation and lithio ethyl acetate[72]. Although widely used in calicene chemistry (see later) this reaction principle − like the Wittig reaction − did not find general application in methylene cyclopropene synthesis.

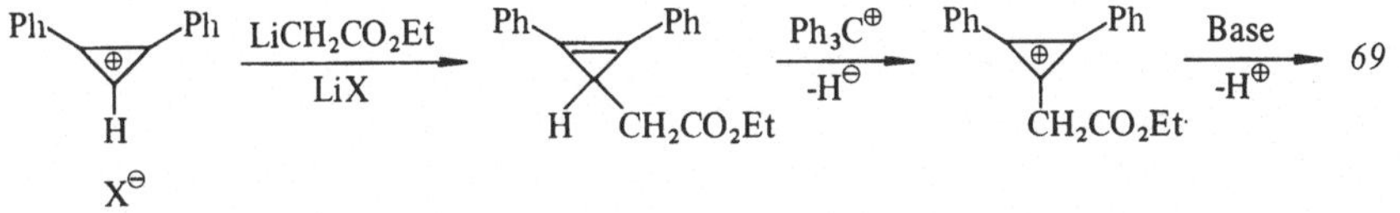

A method of general utility is the "classical" condensation of appropriate reactive methylene compounds (like malononitrile or cyanoacetate) with cyclopropenones[73, 74] − improved by bifunctional acid-base catalysis[67] − in acetic anhydride solution:

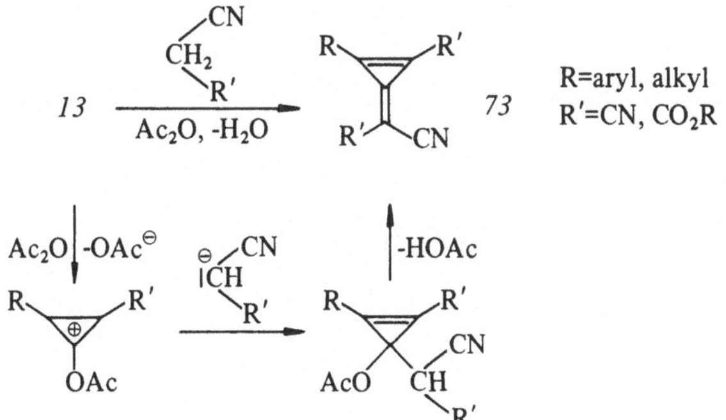

13   73   R=aryl, alkyl
                                          R'=CN, CO₂R

74

13

As was shown for the mechanism of quinocyclopropene formation in acetic anhydride[75] (see p. 20), acylation of the cyclopropenone is reasonable for the primary reaction step, then the O-acyl-cyclopropenium ion *74* forms methylene cyclopropene *73* through addition of the anion of the C-H acidic component and elimination of acetic acid.

The major disadvantages of these methods, *i.e.* relatively small scope and moderate yields (10–25%), are avoided if alkoxy cyclopropenium cations are used for methylene cyclopropene synthesis[76]. These can be prepared easily by alkylation of cyclopropenones with trialkyloxonium tetrafluoroborates[42]. In particular, 3-ethoxy diphenyl cyclopropenium cation (*75*) gave high yields (~80%) of methylene cyclopropenes when reacted with methylene compounds X–CH$_2$–Y (X,Y being CN, COOR, COR, *p*-nitro-phenyl etc.) and a tertiary non-nucleophilic base (preferentially diisopropylethylamine (DIPEA) in a 1 : 1 : 1 ratio[77]. This "cyclopropenylation reaction" can be extended to a large number of X–CH$_2$–Y components and can be carried out on a preparative scale.

The reaction of the anion $^\ominus$|CHXY with *75* shows a marked solvent dependence[60]. Addition of nitriles, *e.g.* CH$_3$CN, suppresses the addition at C^1 – sometimes observed as a side reaction (*77*) – in favor of addition at C^3 resulting in triafulvene formation via *79* – – → *76*. If the amount of base exceeds the 1 : 1 ratio, in some cases, *e.g.* malononitrile and cyanoacetate, triafulvene formation is superseded completely by a ring-opening reaction yielding 1,3-diphenyl-2-alkoxy butadienes *78* probably formed via *77*[60].

Open-chain 1,3-dicarbonyl compounds did not lead to methylene cyclopropenes when reacted with the cation *75* by the DIPEA method. However, 4,4-diacyl triafulvenes *83* can be prepared very easily and in high yields (60–80%) from cations *80* and copper (or zinc) chelates *81* of 1,3-dicarbonyl compounds (known to be capable of electrophilic substitution)[78]. The primarily formed "cyclopropenylated" copper complexes *82* (often isolable because of their insolubility) are readily cleaved to the triafulvenes *83* by means of dilute acid or EDTA.

Triafulvenes derived from cyclic β-dicarbonyl systems like *84/85*[78], *86*[78], and *87*[79] are conveniently prepared from cation *75* by the DIPEA method and in some cases (*87*[80,81], *88*[82], *89*[83]) from diphenyl cyclopropenone by condensation in acetic anhydride.

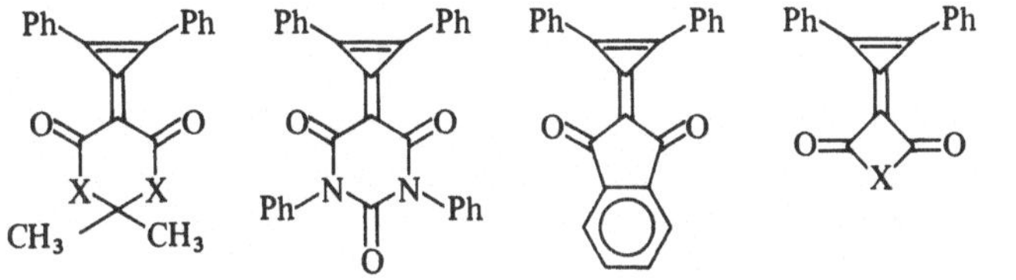

X=CH$_2$: *84*   *86*   *87*   X=CH$_2$: *88*
X=O   : *85*   X=C=CPh$_2$: *89*

The reaction of diphenyl cyclopropenone with aryl malononitriles[75] or aryl cyano acetone[84] unexpectedly gave rise to 4-cyano-4-aryl triafulvenes *90*, as well as the formation of quinocyclopropenes (see later):

90

Special cases of triafulvene formation were found in the base-induced reaction of the nitroso compound *91* with dimethyl fumarate[85], in the thermolysis of tetra-fluorocyclopropene reported to give the perfluorinated triafulvenes *65/66*[68] and in the addition of bis(trifluoromethyl) ketene to bis(p-tolyl) cyclopropenone[66] which gave rise to triafulvene *63* by elimination of CO$_2$:

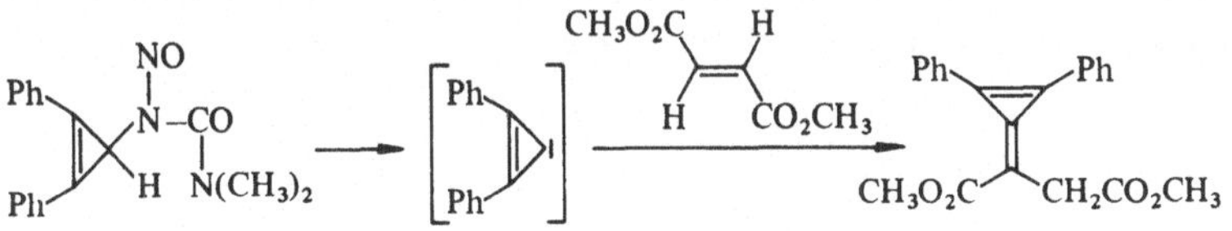

91

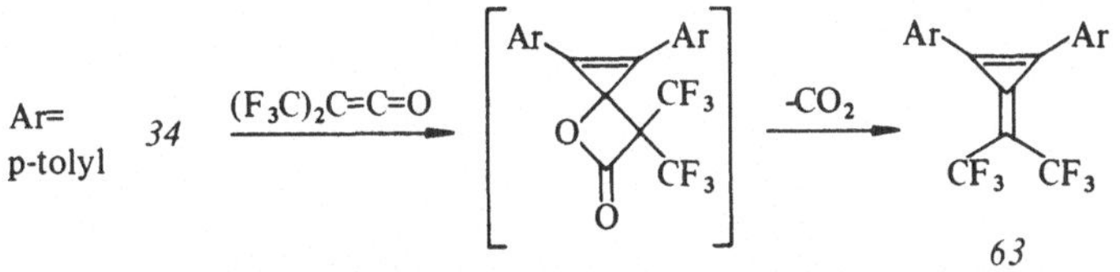

Ar=   *34*   (F$_3$C)$_2$C=C=O
p-tolyl

63

A few examples of vinylogous methylene cyclopropenes are known. Thus, in an interesting reaction mode the spirohexadiene *93*, prepared from dimethyl acetylene and the chlorocarbenoid 92, is claimed to rearrange thermally to the vinylogous triafulvene 94[86]:

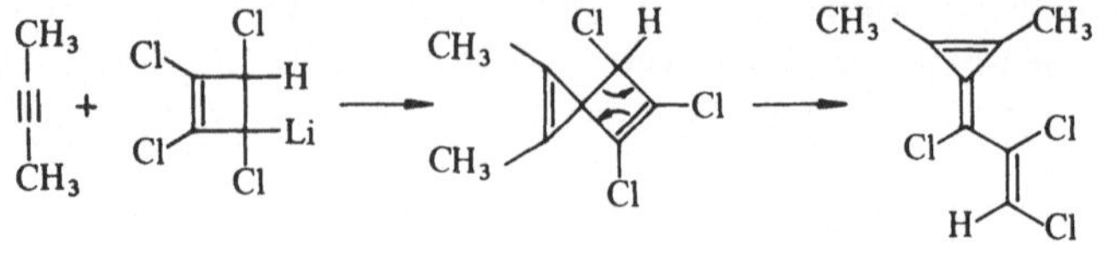

92   93   94

Dicyanomethylene compounds react with ethoxy cation 75 in the presence of DIPEA to form triafulvenes 95 [87, 88]. The same is true for 1,3-bis(dicyanomethylene)-indane and cyanomethylene dimedone, from which the sterically crowded vinylogous triafulvenes 96 and 97 were prepared[88].

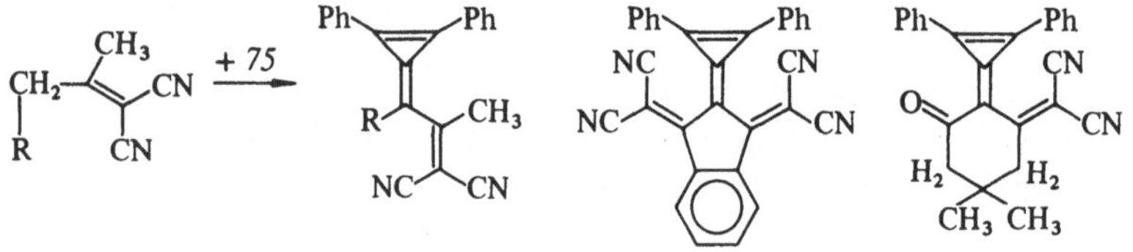

95   96   97

b) Calicenes

Since the chemistry of calicenes has been the subject of former reviews[5, 8], only the principal features for synthesis of calicenes and their mono- and dibenzo derivatives need to be discussed. In general, methods from (a) are used with appropriate variations.

Thus cyclopropenones can be condensed with cyclopentadienes or indenes[89], sometimes under very mild conditions[90], e.g.:

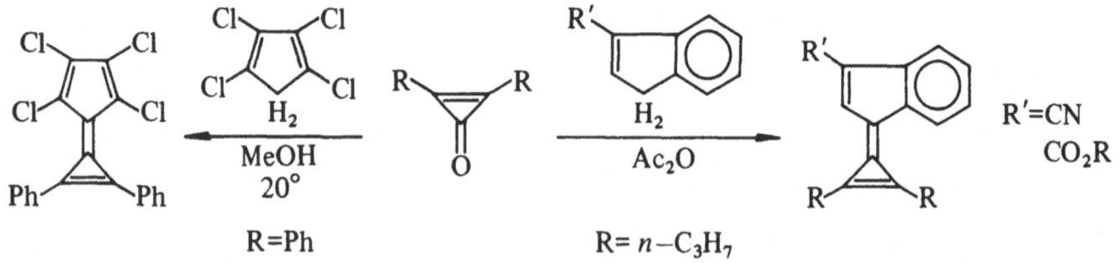

R=Ph   R= n–C$_3$H$_7$

Li, Na, or Grignard compounds of cyclopentadiene, indene, or fluorene add to disubstituted cyclopropenium cations forming cyclopentadienyl cyclopropenes, which can be transformed to calicenes by subsequent hydride abstraction and deprotonation, as shown by the following examples 98[91] and 99[92] (cf. p. 13):

16

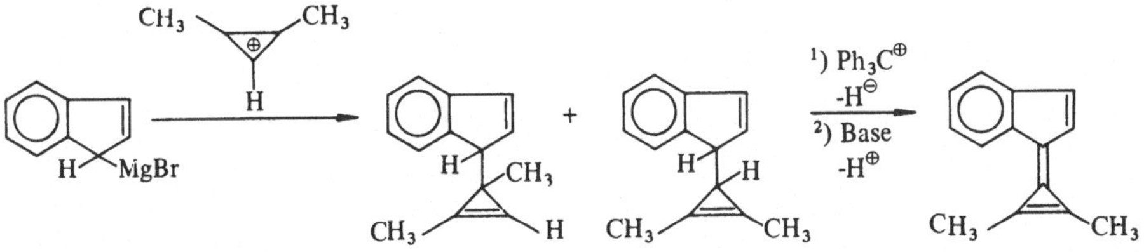

98

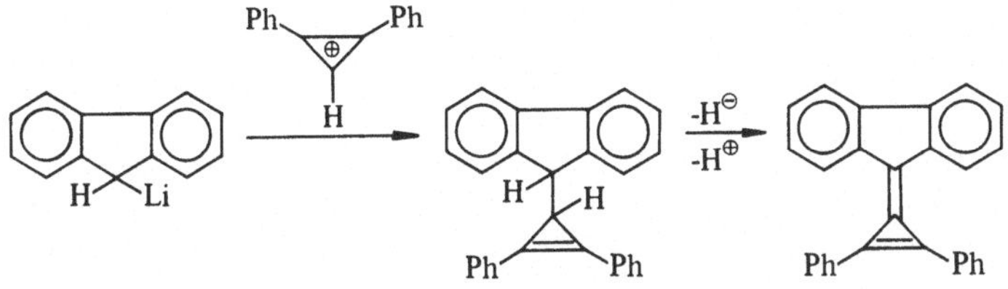

99

Analogously the ethoxy cation *75* was found to be valuable for the synthesis of calicenes when combined with cyclopentadienyl anions[93]. Hexaphenylcalicene (*100*) was prepared by this route from tetraphenylcyclopentadienyl-Li[94, 76] and the highly polar dicyanocalicene *101*[95] from the tetramethylammonium salt of dicyano-cyclopentadiene.

As was recently found by Murata[96], Na-tetrachlorocyclopentadienide yields the calicene *103* only as a minor product when reacted with ethoxy cation *102*; the main reaction consists of an unexpected ring expansion giving rise to dihydropentalenone *104:*

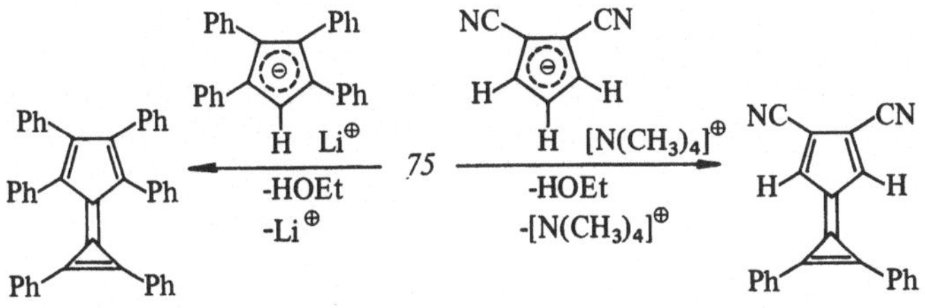

100 *101*

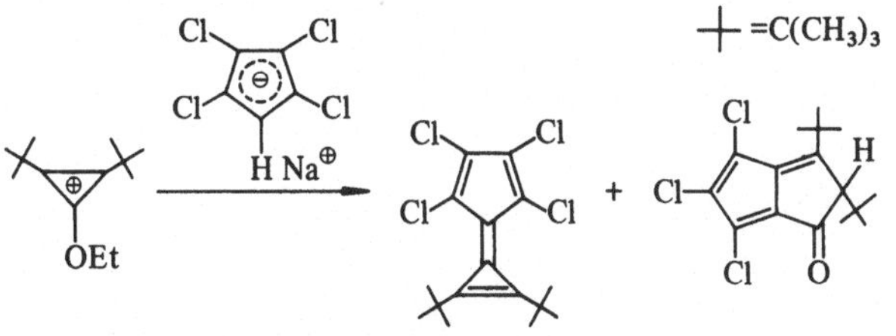

102 *103* *104*

Derivatives of calicene that are unsubstituted at the five-membered ring are unknown with the exception of the cationic complex *105* prepared from ferrocene and 3,3-dichloro-1,2-diphenyl-cyclopropene[97].

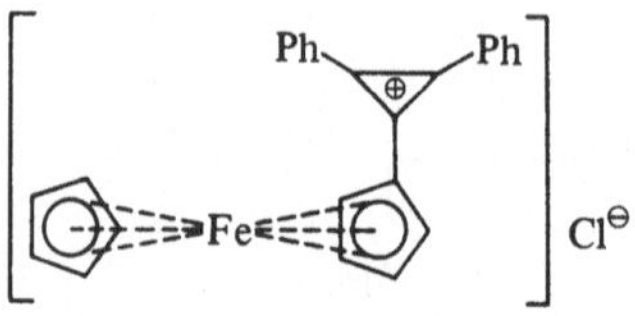

105

c) Quinocyclopropenes

The first stable derivative of methylene cyclopropene was the quinocyclopropene *108* reported in 1964 by Kende[98]; it was prepared from the cyclopropenium cation *106* which underwent pyrolysis and bromination with NBS to the *p*-hydroxy-phenyl substituted cation *107*, which gave quinocyclopropene *108* by deprotonation:

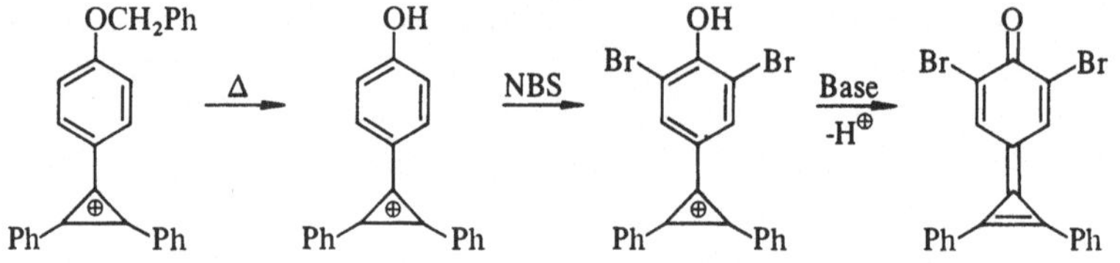

This principle of formation proved to be general for quinocyclopropenes of type *61a* ("phenylogous cyclopropenones"). The required *p*-hydroxy-phenyl cyclopropenium cations were available by electrophilic substitution of phenolic components (preferentially 2,6-disubstituted) and heterosubstituted cyclopropenium cations (*75* and *109*), as the representative examples *110*[99], *111*[76], *112*[99] and *113*[34] show:

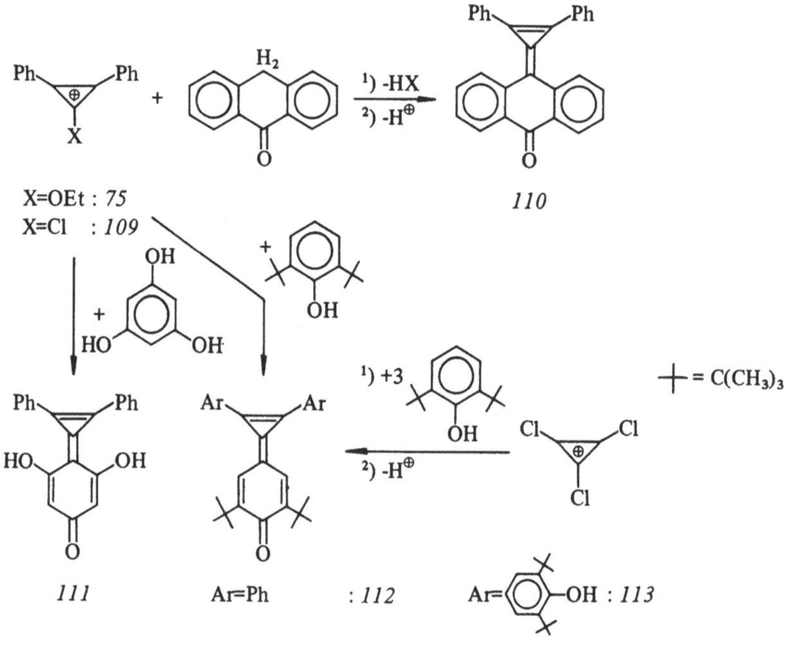

X=OEt : *75*
X=Cl   : *109*

+ = C(CH₃)₃

111   Ar=Ph : *112*   Ar= OH : *113*

Type *61b* of the intensely colored quinocyclopropenes is represented by the dicyanomethylene species *115* and *118* of *p*- and *o*-quinonoid structure. In addition
to the systems *115, 119,* and *122* reported by Gompper[100] a series of *o*- and *p*-quino-
cyclopropenes in the benzene, naphthalene, anthracene, phenanthrene, and fluorene
series (*118–125*) were prepared[75] carrying the bis-(*p*-anisyl)-cyclopropenyl residue,
which brings about a "better" stabilization of the cyclopropenium moiety[101].

Their synthesis is possible from: (a) condensation of arylmalononitriles and
cyclopropenones in acetic anhydride, or (b) thermolysis of $\Delta^{1,2}$-cyclopropene-3-
ethers *114* (adducts of cation *117* and arylmalononitrile anion), as exemplified by
115/116:

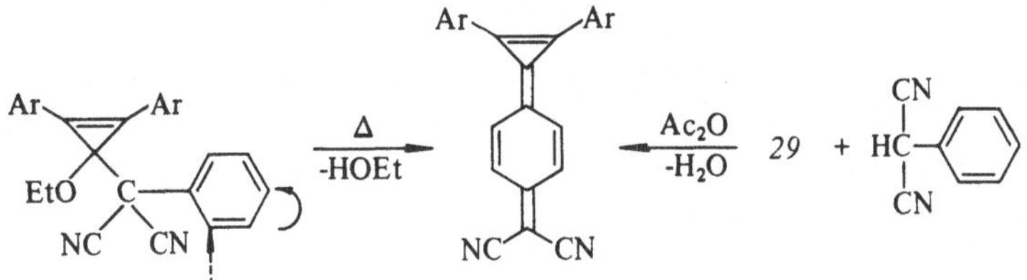

114

115 : Ar=Ph
116 : Ar=An

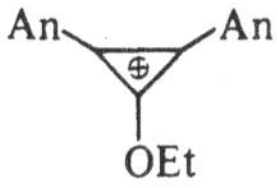

117

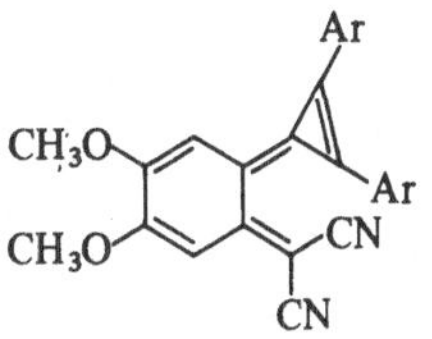

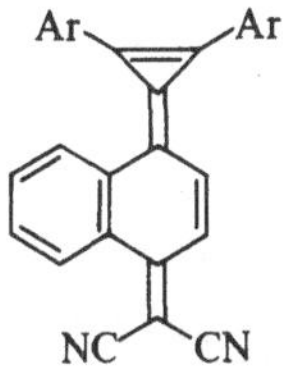

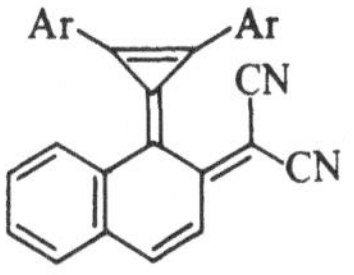

An=*p*-anisyl

118

119 : Ar=Ph
120 : Ar=An

121

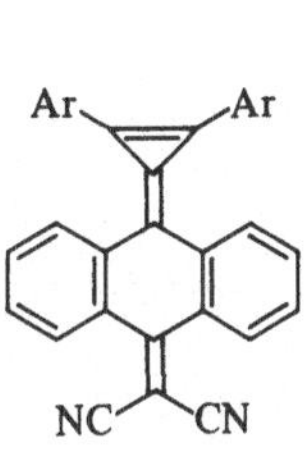

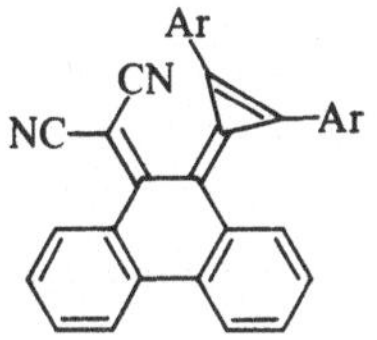

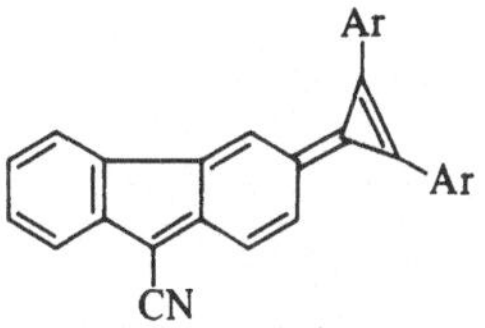

122 : Ar=Ph
123 : Ar=An

124

125

The reaction mechanisms have been clarified in some detail[75]. In method (a) a
complex sequence starts with the acetoxy cyclopropenium ion *126* and the cyclopropenyl acetate *127* and finally leads to adducts *128* containing two moles of arylmalononitrile, which were isolated and shown to be the preferential precursors of
quinocyclopropenes. In method (b) the ambivalent arylmalononitrile anion[102] is
reversibly attacked at the benzylic position at low temperatures, whilst at higher
temperature (after "dissociation" of *114*) attack at the *o*- and *p*-positions of the

19

aromatic nucleus leads to quinocyclopropenes through irreversible elimination of ethanol.

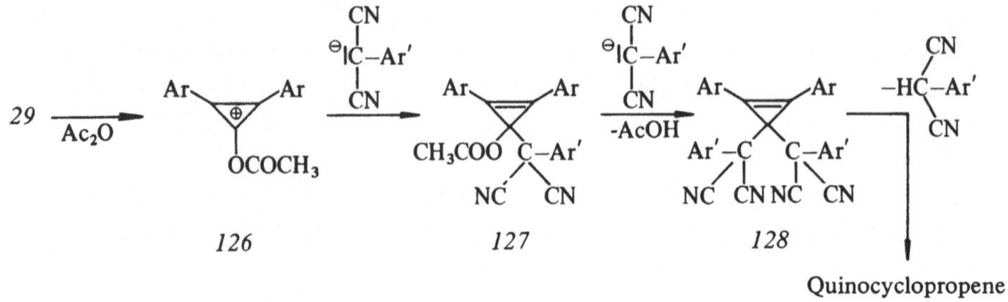

Quinocyclopropene

Related to quinocyclopropenes of type *61a* is the fulvalene dione system *132* synthesized recently from tropolone[103] (and its 1,6-disubstituted derivatives[104]) by means of the ethoxy cation *75*. Of the cations *129/130* formed, primarily *129* is deprotonated to diphenyl heptatriafulvalene-3,4-dione (*132*):

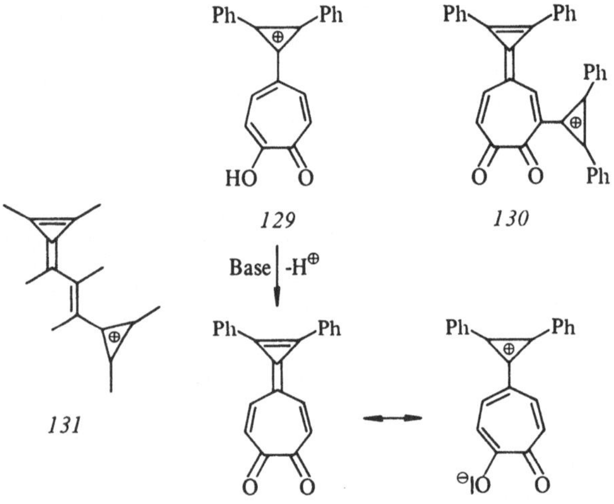

Cation *130* is of interest as a representative of the trimethine cyclopropenylium cyanine system *131*.

Finally, quinonoid phenylogous cyclopropenium immonium cations *133* have been obtained from ethoxy cation *75* and tertiary aromatic amines[61, 76]:

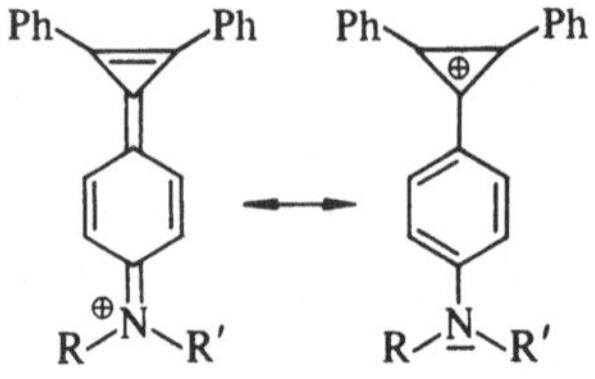

133

d) Cyclopropenium Cyanines

The cyclopropenium system may combine with a large number of donor groups, mainly heterocyclic and carbocyclic systems to give cyclopropenium cyanines. Enamines or ketene acetals of appropriate basicity, e.g. "Fischer Base" (135) or 134, can be easily "cyclopropenylated" by the ethoxy cyclopropenium cation 75 as well as 2- or 4-alkylsubstituted heterocyclic quarternary salts of pyridine, quinoline, and benzothiazol in the presence of a tertiary base[76, 105]:

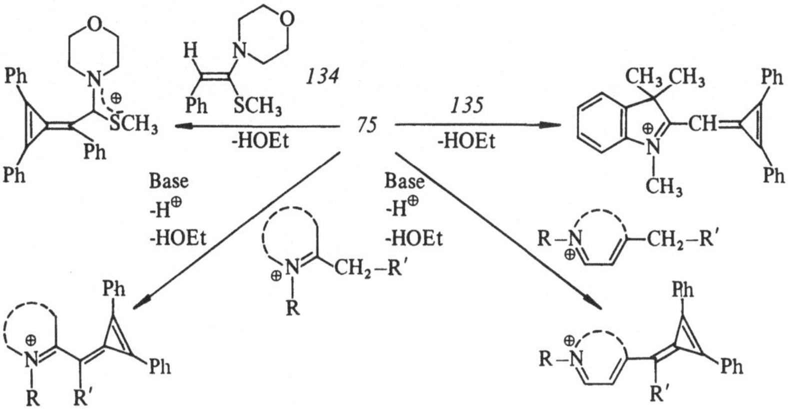

Similiar cyanine types were obtained from the immonium cation 138 and diphenyl cyclopropenone[106, 107], and from 2,3- or 3,4-dimethylindolizinium cations and 75 in the presence of base[105, 108]:

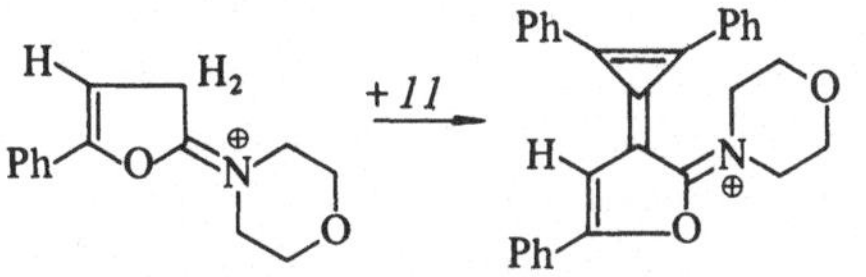

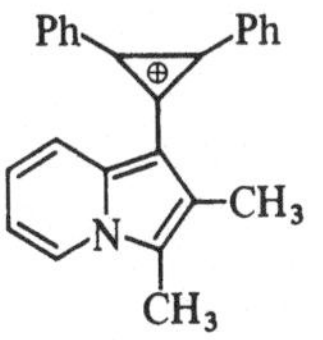

When N-ethylated-α-picoline, γ-picoline, and lepidine were reacted with the ethoxy cation 75, instead of the expected cyanines of type 136 or 137 the products were dicationic species, whose structures were assigned to be bis(diphenylcyclopropenium)-monomethine cyanines, e.g. 142[106]:

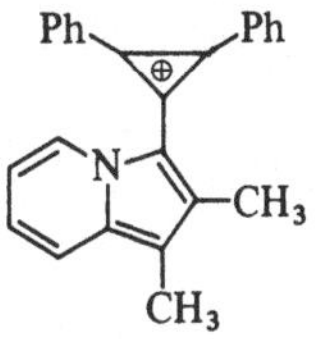

21

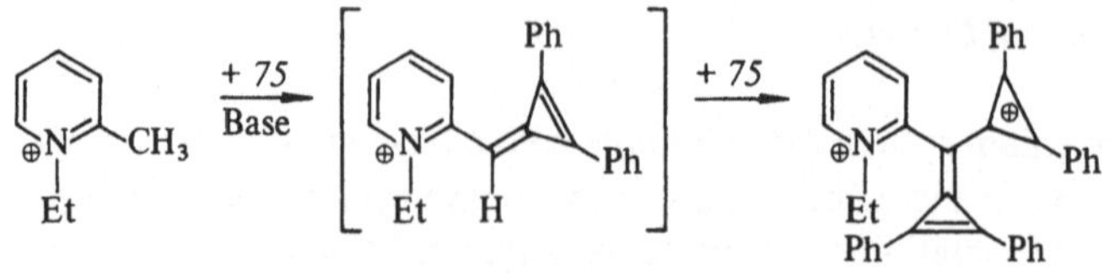

141 *142*

resulting from "cyclopropenylation" of intermediary monocyclopropenium cyanine *141*.

Azulene[106] and 4,6,8-trimethyl azulene[109] readily undergo electrophilic attack by cation *75* at the five-membered ring leading to 1-(1,2-diphenyl-cyclopropenium)-azulenes *143*, which are isoelectronic with indolizinium ions *139/140* and represent cyclic vinylogs of the cyclopropenyl heptafulvenylium system *144*.

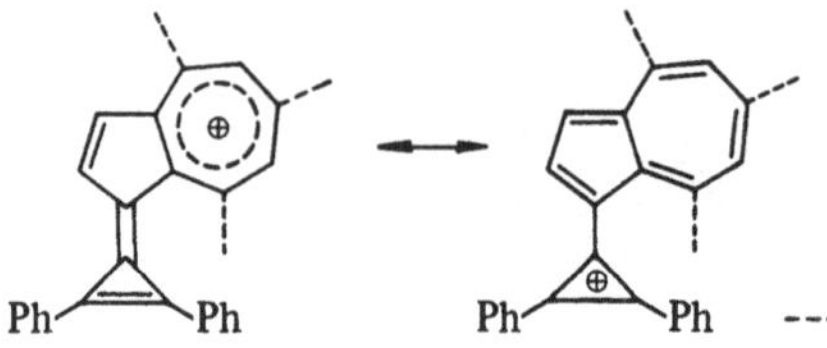

143   ----=CH$_3$

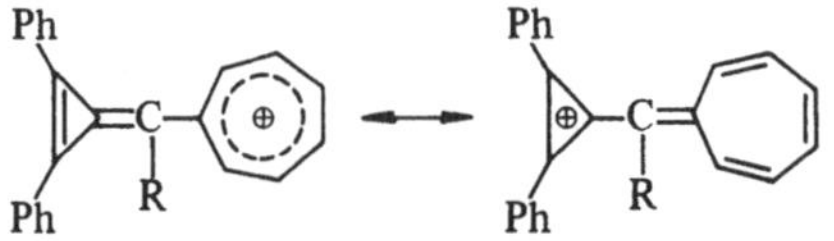

144

The cyanosubstituted diphenylcyclopropenyl heptafulvenylium cation *144* (R = CN) was recently synthesized by Kitahara[110] from 8-cyano heptafulvene and ethoxy cyclopropenium cation *75* and its structure has been proven by X-ray analysis[111].

A final type of cyclopropenium cyanines is found in the azatriapentafulvalenium ions *145*, *146*, and *147/148*, which have been prepared[93, 112, 113] from diphenyl and di-*n*-propyl cyclopropenone or ethoxy cation *75* and indole derivatives as well as from diphenyl methylthio cyclopropenium cation and phenyl-substituted pyrroles[114]:

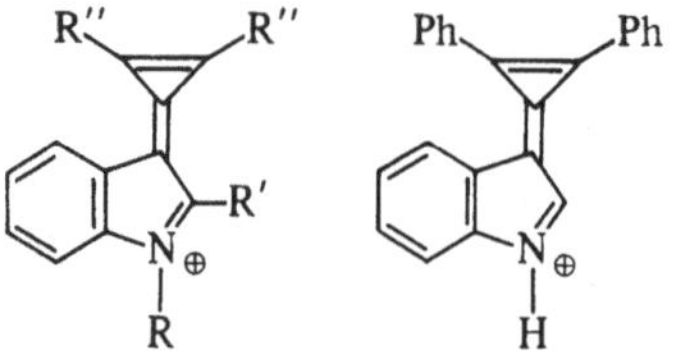

145   *146*

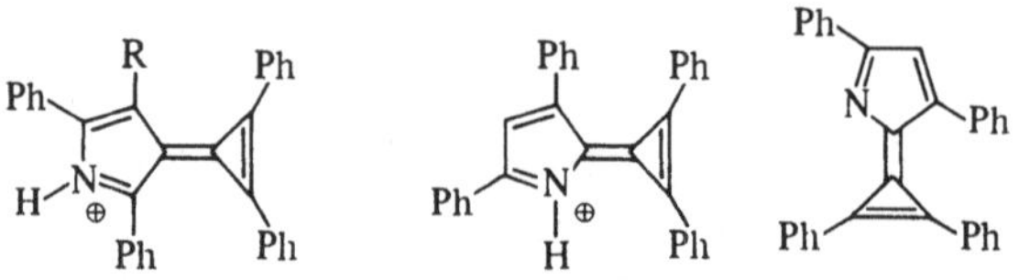

147   *148*   *149*

Deprotonation of *146/147/148* failed to give a defined "azacalicene" derivative, *e.g.* *149*.

Table 2. Preparation of some triafulvenes of type *59*.

	Triafulvene	Yield (%)	Method	Refs.
R = CH$_3$				
	R' = R'' = CN	41	DIPEA	79)
	R' = CN, R'' = COOCH$_3$	34	DIPEA	79)
	R' = R'' = COCH$_3$	22	Cu-chelate	170)
	R', R'' = dimedone	58	DIPEA	59)
	R' = CN, R'' = NO$_2$	43	NH$_4$-salt of NC–CH$_2$–NO$_2$	55)
	R' = CN, R'' = COC$_6$H$_5$	35	DIPEA	219)
	R' = CN, R'' = p–NO$_2$–C$_6$H$_4$	8	DIPEA	219)
	R = n–C$_3$H$_7$	18	Acetic anhydride	74)
	R' = R'' = CN			
	R = –(CH$_2$)$_5$–	58	DIPEA	79)
	R' = R'' = CN			
Ar = C$_6$H$_5$				
	R' = R'' = CN	85	DIPEA	77)
		24	Acetic anhydride + alanine	67)
	R' = CN, R'' = COOCH$_3$	82	DIPEA	77)
	R' = CN, R'' = COOC$_2$H$_5$	70	DIPEA	253)
		15	Acetic anhydride + alanine	67)
	R' = CN, R'' = COC$_6$H$_5$	66	DIPEA	298)
	R' = R'' = COCH$_3$	80	Cu-chelate	78)
	R' = R'' = COC$_6$H$_5$	80	Cu-chelate	300)
	R' = COC$_6$H$_5$, R' = CO$_2$C$_2$H$_5$	62	Cu-chelate	78)
	R' = COCH$_3$, R'' = CONHC$_6$H$_5$	78	Cu-chelate	78)
	R' = COC$_6$H$_5$, R'' = CHO	74	Cu-chelate	78)
	R', R'' = dimedone	64	DIPEA	77)
	R', R'' = meldrum acid	53	DIPEA	77)
	R', R'' = indane dione	64	Acetic anhydride	80)
	R' = H, R'' = –CPh=C(CN)$_2$	24	DIPEA	88)
	R' = COOC$_2$H$_5$ R'' = –C=C(CN)$_2$ | CH$_3$	95	DIPEA	88)
	R' = CN, R'' = NO$_2$	61	NH$_4$-salt of NC–CH$_2$–NO$_2$	55)
	Ar = p–CH$_3$–C$_6$H$_4$	59	*34* + (CF$_3$)$_2$C=C=O	66)
	R' = R'' = CF$_3$			

23

Table 2 (continued)

	Triafulvene	Yield (%)	Method	Refs.
R = CH$_3$				
	R' = R'' = COCH$_3$	74	Cu-chelate	61)
	R' = COCH$_3$, R'' = CONHC$_6$H$_5$	96	Cu-chelate	61)
	R' = CN, R'' = NO$_2$	96	NH$_4$-salt of NC−CH$_2$−NO$_2$	55)
	R' = R'' = CN	19	DIPEA	55)
	R', R'' = dimedone	65	DIPEA	55)
R = C(CH$_3$)$_3$		27	DIPEA	63)
R' = R'' = CN				
R = H				
	R' = R'' = COCH$_3$	16	Cu-chelate	55)
	R' = COCH$_3$, R'' = COC$_6$H$_5$	27	Without base	55)

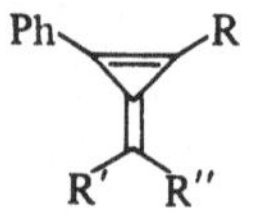

3. Derivatives of Cyclopropenone ("Heterotriafulvenes")

The synthesis of cyclopropenone imines *3* has been accomplished by several methods. Thus aromatic amines, *e.g.* *p*-nitraniline, can be reacted either with diphenyl cyclopropenone in HCl/ethanol or with the ethoxy cation *75* forming the immonium cation *150*, which is deprotonated by tertiary bases to the N-(*p*-nitro-phenyl)-imine *151*[115]:

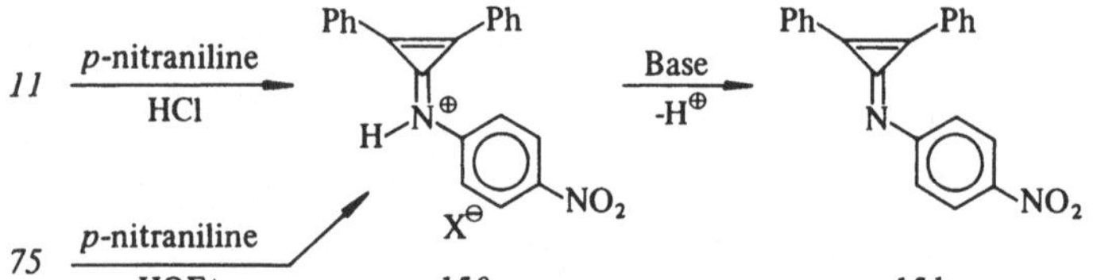

$$150$$
$$(X=Cl, BF_4)$$

Diphenyl cyclopropenone reacts readily with isocyanates activated by *p*-toluene-sulphonyl, trichloroacetyl, and chlorosulphonyl[116] or benzenesulphonyl[117] groups giving rise to cyclopropenone imines *152* and carbon dioxide:

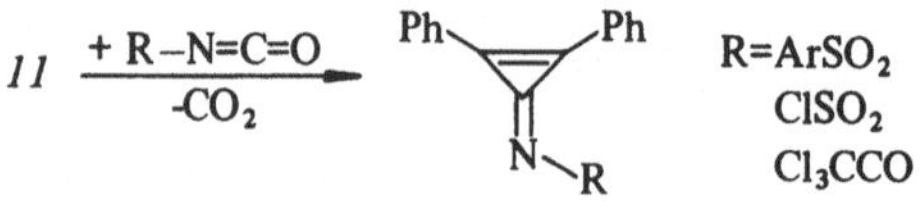

An elegant method for the preparation of some cyclopropenone imines reported by Krebs[118] is the (1 + 2) cycloaddition of isonitriles (as divalent carbon species) to activated triple bond of ynamines and certain cycloalkynes, *e.g.*:

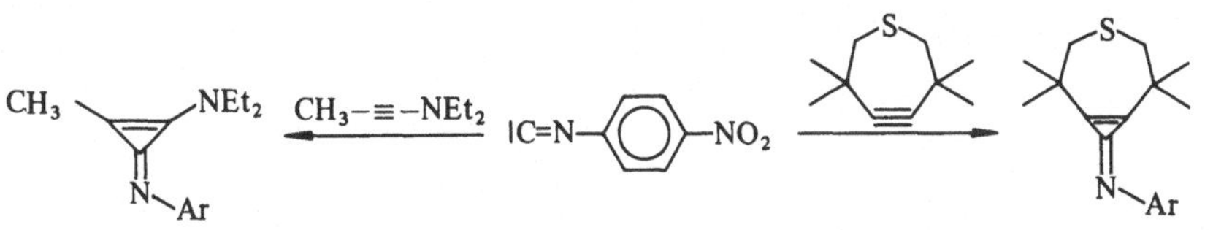

Ar=p-nitrophenyl

The highly stable cyclopropenium immonium cations *153* are formed from the ethoxy cyclopropenium cation *75* with primary and secondary aliphatic or aromatic amines[42, 119]:

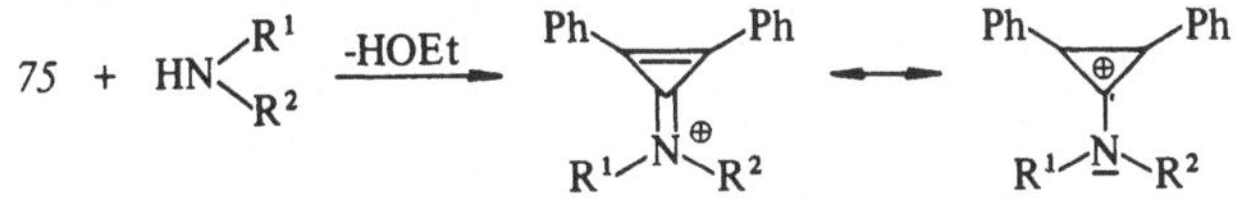

153

Diphenylcyclopropene thione (*156*) was prepared[115, 120] from 3,3-dichloro-1,2-diphenyl cyclopropene (*154*) by reaction with thioacetic acid, since transformation of the carbonyl function of diphenyl cyclopropenone with P_4S_{10}[121] was complicated by ring expansion to the trithione *155*[122]. In a useful recent thioketone synthesis[123] *156* was obtained directly from diphenyl cyclopropenone in a quantitative yield by simultaneous treatment with HCl and H_2S.

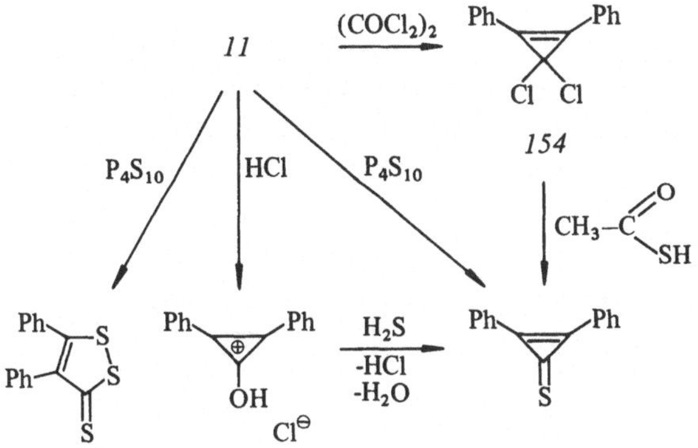

155   *156*

Of the further functional derivatives of cyclopropenones in the diphenyl series, the oxime[115, 121] and several hydrazones[115] (*e.g. 158/159*), and azines (*e.g. 160*[115]), are easily available from the ethoxy cation *75* and hydroxylamine, hydrazines, and hydrazones, respectively. Sometimes oximation of cyclopropenones produces unexpected results (see later and Ref.[42]).

Whilst anhydrous hydrazine is reported[124] to give products of nucleophilic ring opening and cyclization (*162/163*) with diphenyl cyclopropenone, hydrazine hydrochloride yielded the azine *161*[125].

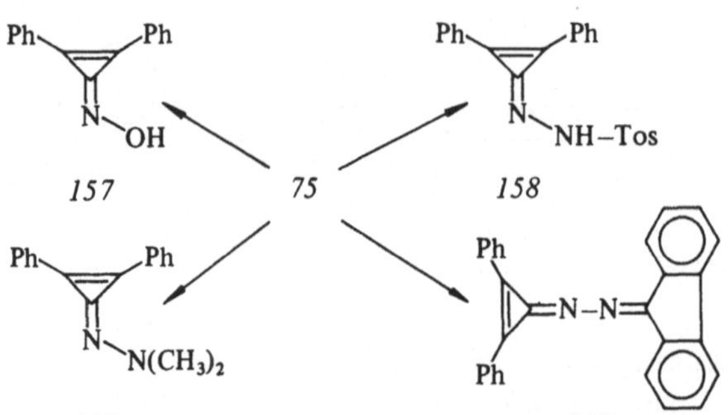

157   75   158

159   160

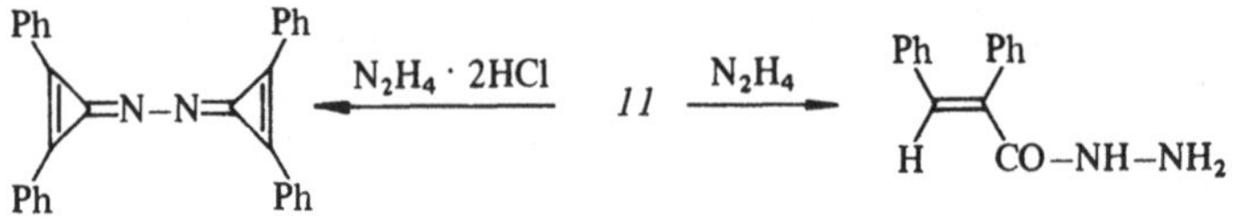

161            $N_2H_4 \cdot 2HCl$ 11 N_2H_4 162

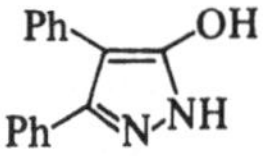

163

4. Benzocyclopropenones

Although benzocyclopropenes *164* have been isolated[126] and were found to be moderately stable, derivatives of the more strained but electronically more stabilized benzotriafulvene system *165* have not yet been synthesized. However, benzocyclopropenones have been shown to be reactive intermediates in several reactions.

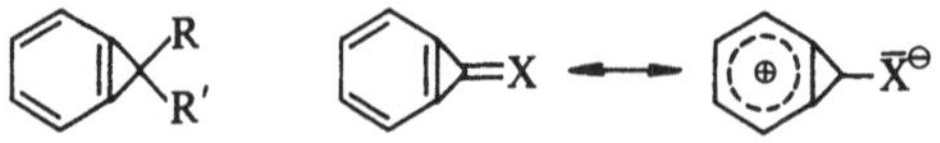

164   165

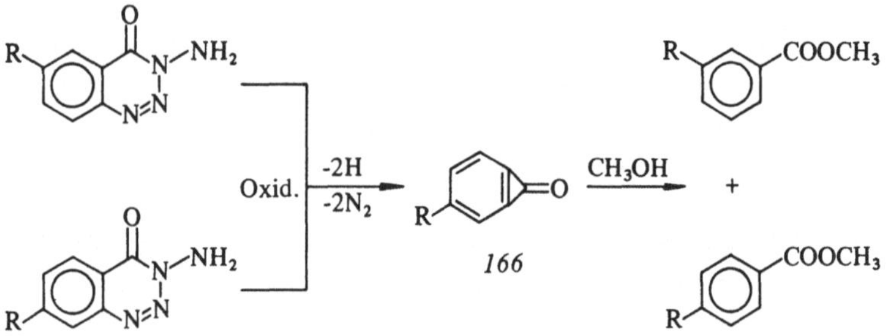

Oxid. $\dfrac{-2H}{-2N_2}$ 166 CH_3OH

Thus oxidation of both 6- and 7-substituted 3-amino-benzo-1,2,3-triazin-4-ones with lead tetraacetate in methanol produced a mixture of *p*- and *m*-substituted benzo-ates, clearly indicating that a symmetrical intermediate, *i.e.* benzocyclopropenone (*166*) was formed and underwent ring opening by attack of solvent[127].

Only the appearance of the "rearranged" product is significant for benzocyclopropenone intermediacy, since the indazolone *168* was detected and trapped by tetracyclone via *170* in the oxidation of *167*, which was shown to yield only "unrearranged" ester *169* by nucleophilic cleavage by the solvent; thus *168* cannot be a precursor of benzocyclopropenone *171*.

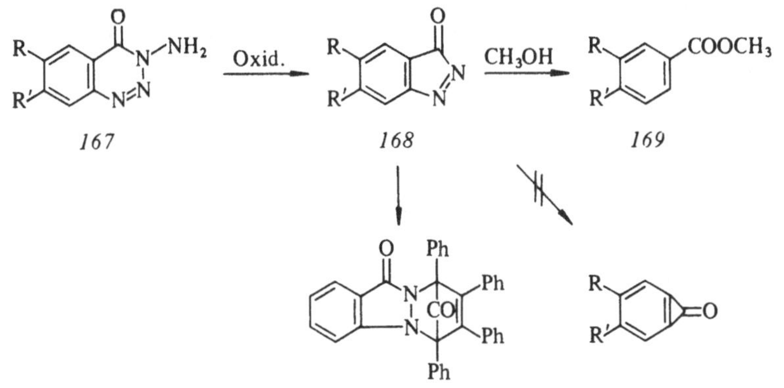

The intervention of an intermediate with the symmetry of a benzocyclopropenone (*166*, R = Cl) is also demanded by the formation of methyl *p*-chlorobenzoate in the photolysis of the lithio derivative *173* of the chlorosubstituted 3-*p*-tolylsulphonylamino-benzo-1,2,3-triazin-4-one[128]:

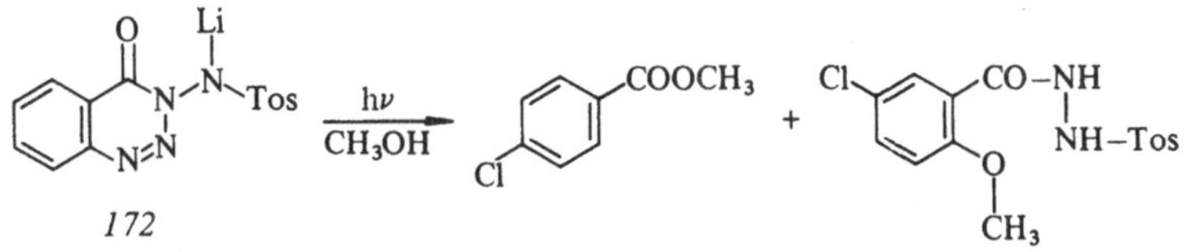

Attempts were reported[119, 129] to synthesize the ortho-linked diphenyl cyclopropenone "phenanthreno cyclopropenone" *173* by dehydrohalogenation of the dibromo derivative *174* of dibenzo cyclohepta-1,3-diene-6-one. The only product isolated was the anhydride *175* of phenanthrene-9-carboxilic acid, which was shown not to arise from *173*[129].

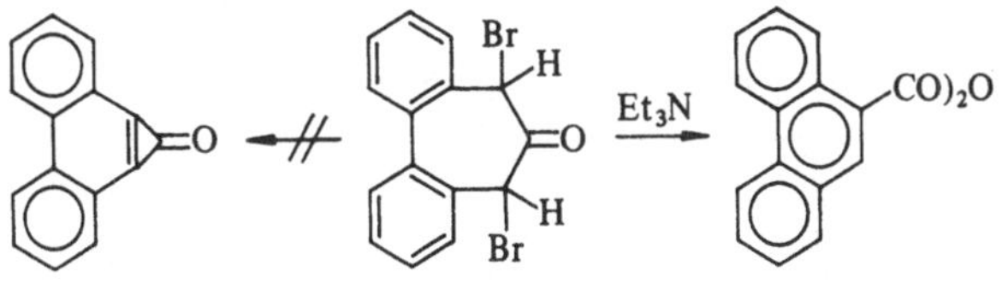

27

A benzocyclopropenone intermediate *177* may account for the formation of terphenic acid *178* on hydrolytic decomposition of benzocyclopropenium cation *176*[130a]. Likewise, the intermediacy of benzocyclopropenone (*166*, R = H) is claimed from spectroscopic evidence in the low-temperature photolysis of benzocyclobutene dione leading to benzyne[130b].

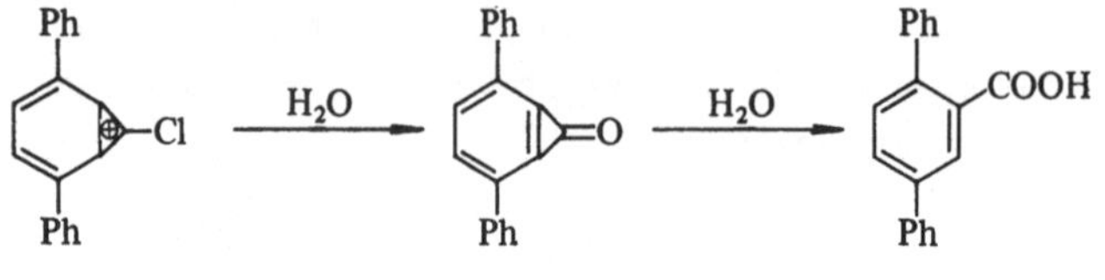

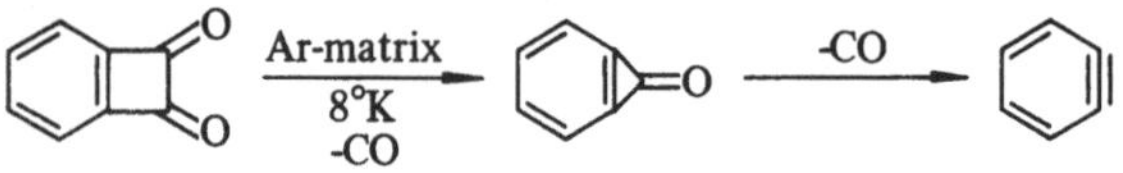

III. Structural Criteria for Cyclopropenones and Triafulvenes

1. Basicity and Dipole Moment of Cyclopropenones and Triafulvenes

a) Basicity

Cyclopropenones can be isolated from the organic phase either by extraction or by precipitation with strong acid[42, 43] in the form of stable, often well-defined protonation products, the hydroxy cyclopropenium salts *179*:

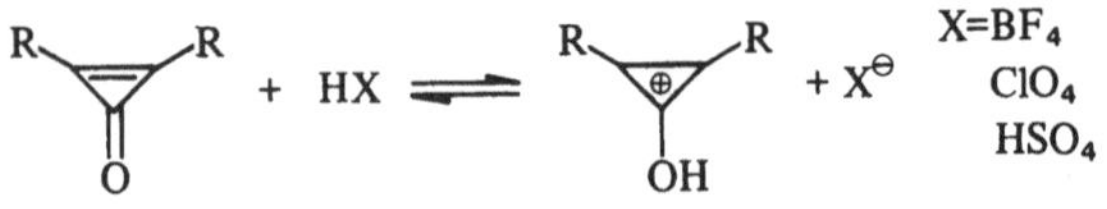

179

The basicity of cyclopropenones was determined by examining the disappearance of typical IR-absorptions (see p. 38) on protonation or the change in the ^{1}H-NMR chemical shifts of three-ring substituents as a function of H_0 in solutions of various acidity. From these measurements half-protonation was found to occur as follows:

Table 3. Basicity of cyclopropenones

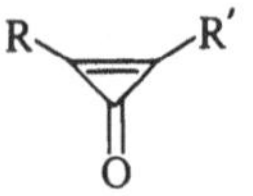 R = R′ = H	−5.2[28]
R = H, R′ = CH$_3$	−3.5[23]
R = R′ = CH$_3$	−2.3[23]
R = R′ = n−C$_3$H$_7$	−1.9[43]

28

Table 3. (continued)

	R = R' = cyclopropyl	-1.2[57)]
	R = R' = C_6H_5	-2.5[42)]
For comparison:		
	Tropone	-0.4[4)]
	Eucarvone	-4.9[131)]

The above values can only represent a rough measure of basicity, since it has been found[23, 43)] that cyclopropenones do not behave as Hammett bases. A refined treatment minimizing the NMR effects extraneous to the protonation of interest gave H_0 values of -5.0 for methyl cyclopropenone and -1.5 for dimethyl cyclopropenone[23)].

Cyclopropenones are markedly more basic than other α,β-unsaturated compounds[131)]; tropone, however, exhibits higher basicity than cyclopropenones due to its half-protonation value of -0.4. As shown in Table 4, basicity of cyclopropenones is increased by alkyl substituents relative to phenyl substitution. This is also demonstrated by the observation[132)] that di-n-propyl cyclopropenone is extracted by 12N HCl from CCl_4 solution, but diphenyl cyclopropenone is not.

This effect might be due to the "better" stabilization of the protonated species *179* by electron-donating groups, which is parallelled in the increase of pK_{R^+} of cyclopropenium cations when going from triphenyl to trialkyl substitution[101)].

Similiar estimations of the basicity of triafulvenes have not yet been performed. It might be of qualitative interest that 1,2-diphenyl-4,4-diacetyl triafulvene (*180*) also forms a stable hydrofluoroborate[78)] (*181*, X = BF_4) as it is extracted by 12N HCl from $CHCl_3$ solution, while analogous behavior is not observed for dimethyl and diphenyl 4,4-dicyano triafulvene (*182/64*).[79)]

180   181   R=CH₃ : *182*
                                                     R=Ph :   *64*

Further aspects of triafulvene protonation are given on p. 90.

b) Dipole Moments

As shown in Table 4, high dipole moments have been found to be characteristic of cyclopropenones and triafulvenes of various types.

As comparison with Table 3 shows, dipole moments and basicities of cyclopropenones are not correlated: di-n-propyl and di-cyclopropyl cyclopropenone are stronger bases than diphenyl cyclopropenone; the latter, however, possesses the higher dipole moment.

Table 4. Dipole moments of cyclopropenones and triafulvenes (given in D)

(cyclopropenone structure)	R = H 4.39[133] $\xrightarrow{R = H}$	calculated values:
	R = n-C$_3$H$_7$ 4.78[43]	3.58[134] (*ab initio*)
	R = cyclopropyl 4.58[57]	4.67[137a] (*ab initio*)
	R = $-$(CH$_2$)$_5-$ 4.66[43]	4.52[135] (CNDO/2)
	R = C$_6$H$_5$ 5.08[17]	5.09, 3.73[136] (CNDO/2)
	5.14[42] $\xrightarrow{R = Ph}$	calculated values:
		4.43[137b]
(Ph cyclopropenethione structure)	5.8[121]	6.63[138a] (MO–CI)
		5.13[135] (CNDO/2)

(triafulvene structure, Ar, Ar, R, R')

Ar = p-tolyl, R = R' = CF$_3$ 7.42[66]
Ar = phenyl, R = R' = CN   7.9[67]
Ar = phenyl, R = CN, R' = COOC$_2$H$_5$ 5.9[67]

(Ph structure)	(Ph structure)	(Ph structure)	(R structure)
9.4[99]	6.3[94]	14.3[95]	R = n-C$_3$H$_7$ 7.56[139]
			R = Ph 7.97[140]
			8.10[141]

For comparison:

CH$_3$–CO–CH$_3$ 2.85[135]

C$_6$H$_5$–CO–C$_6$H$_5$ 3.0[4]

(cyclohexenone structure with CH$_3$ groups)   4.0[142]

(tropone structure) =O   4.30[4]

(CH$_3$)$_3$\overset{\oplus}{N}$–$\overset{\ominus}{O}$ 5.03[4]

$\overset{\oplus}{}$----$\overset{\ominus}{}$ μ=4.8 D
1Å

Since the dipole moments of cyclopropenones are enlarged with respect to simple ketones and compare to other polar systems, *e.g.* trimethylamine oxide in Table 4, there seems to be evidence for considerable charge separation in the carbonyl group, which was expressed in terms of a "cyclopropenium oxide" contribution to the ground state.

To a first approximation[143] the dipole moment of cyclopropenone can be considered as an additive combination of the moments of structures *a* and *b*, which were estimated from group moments

μ_a 2.22   (structure *a*)   0.45 / 2.67   (structure *b*)   1.5   10.46   11.96 μ_b

a   *b*

involved in a and b: cyclopropene (0.45), cyclopropanone (2.67[144]), and the completely charge-separated cyclopropenium oxide structure [1.5 (C—O) + 10.46]. According to equations

$$\mu = \mu_a (1 - x) + \mu_b \cdot x; \quad x = \frac{\mu - \mu_a}{\mu_b - \mu_a} = 0.23$$

using the known[133] dipole moment of cyclopropenone (4.39 D) the contribution x of the zwitterionic structure b to the groundstate hybrid amounts to 23%.

In contrast to this estimation, Tobey[58] argued that the enhanced moments observed in cyclopropenones merely reflect the increased distance between the centers of negative (oxygen atom) and positive (middle of three-ring) charge compared to the carbonyl group with a dipole moment of 2.8 ± 0.2 D. This hypothesis is questioned by the results of Ammon[135] discussed later.

Finally, calculations of dipole moments for cyclopropenone, dimethyl cyclopropenone[136], and diphenyl cyclopropenone (Table 4) have been reported. The earlier results markedly suffer from the uncertainties of cyclopropenone bond lengths which have only recently been established (see following chapter).

2. Molecular and Electronic Structure of Cyclopropenones and Triafulvenes

a) The Structure of Cyclopropenone

Breslow et al.[133] investigated the microwave spectrum of cyclopropenone and determined data for bond lengths, bond angles, dipole moment (4.39 D from the molecular Stark effect), and magnetic susceptibility anisotropy (Δ_χ) as seen in Table 5 in comparison with cyclopropene[53].
It is interesting to note that the $C^1{=}C^2$ distances in cyclopropenone and cyclopropene are nearly identical, whilst $C^{1\,(2)}{-}C^3$ is shorter in cyclopropenone. The opposite trends were observed for cyclopropanone-cyclopropane single-bond relationships.

An important aspect is derived[133] from the Δ_χ-value currently substantiated as a criterion for the extent of ring electronic delocalization[145]. Going from propene ($\Delta_\chi = -6.4$) to cyclopropene a diamagnetic ring current seems to be established which should be increased by replacing CH_2 by the electron-withdrawing C=O group. Although this leaves the three-ring with the diamagnetic $4n + 2$ configuration of π-electrons, no appreciable increase in Δ_χ for cyclopropenone is observed.

This lack of a major diamagnetism may be attributed to two factors. First even in the completely delocalized cyclopropenium cation the diamagnetic ring current effects are small and in the range of only 25% of those in benzene as concluded from NMR data[146].

Second, cyclopropenone is apparently not completely "delocalized" and might be regarded as a resonance hybrid 183 of unequivocal contributions $a-d$ which differ in energy and are not simply "mixed" with equal weight, as implied in the "cyclopropenium oxide" symbolism hitherto used.

31

Table 5. Structural data of cyclopropenone compared to cyclopropene (obtained from microwave spectra)

		Cyclopropenone	Cyclopropene
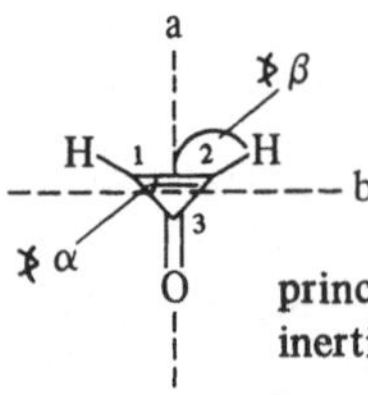 principal inertial axes Symmetry: C_{2v}	Distances (Å)		
	C=O	1.212	----
	$C^1=C^2$	1.302	1.300
	C^3-C^1 (2)	1.412	1.515
	$C-H_{vinyl}$	1.097	1.070
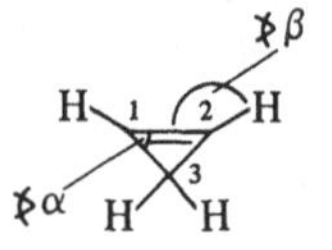	Angles		
	$\sphericalangle\,\alpha$	62° 33'	50° 48'
	$\sphericalangle\,\beta$	144° 55'	149° 55'
Magnetic susceptibility anisotropy	$\underline{\Delta}_\chi$	-17.8 ± 1.0	-17.0 ± 0.5

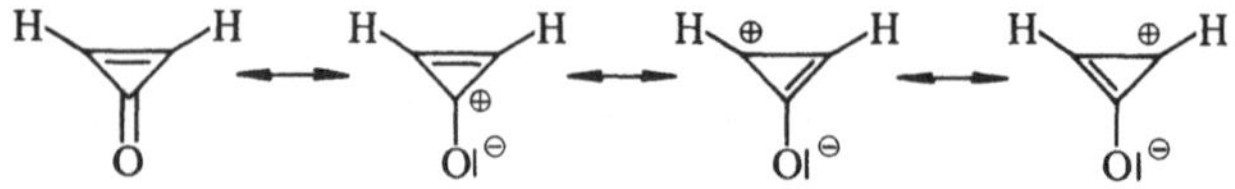

a   b   c   d   183

In addition to the determined molecular geometry this is supported by ^{13}C–NMR signals at -158.3 ppm (C^1/C^2) and -155.1 ppm (C^3) relative to TMS; for comparison, cyclopropenium cation has its ^{13}C–NMR signal at -174 ppm[147]).

Thus "aromaticity" in cyclopropenone cannot be detected by the magnetic criterion although it is suggested by other chemical and physical properties. UV, IR, and ^{1}H–NMR data of cyclopropenone are summarized in Table 6, but are discussed in later chapters.

b) Structure Determination of Substituted Cyclopropenones and Triafulvenes

The molecular structures of diphenyl cyclopropenone (anhydrous[148] and as hydrate[135, 149], diphenylcyclopropene thione[150], 1,2-diphenyl-[135], and 1,2-dimethyl-[151a]) 4,4-dicyano triafulvene, 1,2-di-(p-tolyl)-4,4-di-(trifluoromethyl) triafulvene[151b], 5,6-diphenyl-[152], and 5,6-di-n-propyl-[153] 1,2,3,4-tetrachlorocalicene and 8-cyano-8-(diphenylcyclopropenyl)-heptafulvenylium tetrafluoroborate[111] have been determined by X-ray analysis[154].

Table 6. UV, IR, and ^{1}H–NMR data of cyclopropenone

UV:	276 nm (log = 1.49)	$(n \rightarrow \pi^*)$
	below 190 nm	$(\pi \rightarrow \pi^*)$
IR:	cyclopropenone-H_2	1864, 1833, 1480 cm^{-1}
	cyclopropenone-D_2	1858, 1779 cm^{-1}
	cyclopropenone-^{18}O	1834, 1817 cm^{-1}
	cyclopropenone-$D_2-^{18}O$	1858, 1764 cm^{-1}
^{1}H–NMR:	vinyl-H	1.0 τ (D$_2$O), 0.92 (CH$_3$NO$_2$)
		0.4 τ (H$_2$SO$_4$)

J_{13C-H} = 217 Hz, J_{H-H} = 3.9 Hz (CDCl$_3$)
$\phantom{J_{13C-H}}$ = 250 Hz, $\phantom{J_{H-H}}$ = 1.3 Hz (H$_2$SO$_4$)

X=O: *11*
X=S: *156*

R=CH$_3$: *182*
R=Ph : *64*

63

R=n−C$_3$H$_7$: *184*
R=Ph : *185*

144 (R=CN)

In Table 7 the parameters relevant to a structural discussion of the three-ring moiety in these compounds are listed.

Comparison of the observed bond distances with distances known for appropriate three-ring reference systems (cyclopropanone: C=O 1.191 Å[155], cyclopropene: C=C 1.30 Å[53], triisopropylidene cyclopropane: C–C 1.44 Å[156]) have led to the following conclusions.

(1) The carbonyl distances in diphenyl cyclopropenone and cyclopropenone (1.225/1.212 Å) are larger than in cyclopropanone and indicate enhanced single-bond character. The same is true for the C=S bond in the thione *156* (1.63 Å) compared to the C=S distance in thioketones (1.56 Å[157]).

(2) The C^3-$C^{1(2)}$ distances in the substituted species are not markedly affected by the exocyclic group X and agree well with unsubstituted cyclopropenone (Table 3). The 0.05 Å difference in C=C between cyclopropenone and its diphenyl derivative may be associated with an (unspecified) effect of the phenyl substituents or of the hydrate water molecule.

(3) The mean of C^1-C^2 and C^3-$C^{1(2)}$ distances is in all cases near to the 1.373 Å three-ring length in triphenyl cyclopropenium perchlorate[158] and the 1.363 Å distance in tris (dimethylamino) cyclopropenium perchlorate[159]. Likewise, the $C^{1(2)}$-

33

Table 7. Bond distances (Å) and bond angles (°) in the three-ring of some substituted cyclopropenones and triafulvenes

Compound	$C^1=C^2$	$C^3-C^{1(2)}$ (average)	C=X	$C^{1(2)}$–Ph	$\angle\alpha$ (average)	$\angle\beta$	$\angle\gamma$	Ph-twist	Refs.
Diphenyl Cyclopropenone (11)	1.349	1.417	1.225 (C=O)	1.447	61.6	150.6	56.9	2.2	148)
Diphenyl cyclopropenone (hydrate)	1.354	1.409	1.226 (C=O)	1.452	61.3	149.3	57.4	6.2	135)
	1.36	1.395	1.23 (C=O)	1.44	61	149	58	–	149)
Diphenylcyclopropene thione (156)	1.338	1.403	1.630 (C=S)	1.440	61.5	152	57.5	4	150)
1,2-Diphenyl-4,4-dicyano triafulvene (64)	1.344	1.398	1.367 (C^3-C^4)	1.444	61.3	151.3	57.5	5.9	135)
1,2-Dimethyl-4,4-dicyano triafulvene (182)	1.327	1.393	1.367 (C^3-C^4)	–	61.6	–	56.9	–	151a)
1,2-Di(p-tolyl)-4,4-di(trifluoromethyl) triafulvene (63)	1.342	1.417	1.357 (C^3-C^4)	1.445	61.7	149.0	56.5	4.06 5.20	151b)
5,6-Diphenyl-tetrachloro-calicene (185)	1.349	1.413	1.357 (C^3-C^4)	1.460 1.433	61.5	146.3	57.0	45 0	152)
5,6-Di-n-propyl-tetrachloro-calicene (184)	1.320	1.390	1.370 (C^3-C^4)	–	61.7	151.0	56.7	–	153)
Heptafulvenylium cation 144 (R=CN)	1.353	1.400	1.380 (C^3-C^4)	1.431 1.443	60.9 57.8	–	61.3	27.0 6.5	111)

phenyl bonds agree well with triphenyl cyclopropenium cation (average 1.436 Å) and are considerably shorter than the accepted $C(sp^2)$–$C(sp^2)$ bond length (1.48 Å[150]). However, "twisting" of phenyl rings in cyclopropenium cation amounts to 7.6°, 12,1°, and 21.2°, respectively[158].

(4) It is believed[135, 148] that the small differences between C–C and C=C in the above cyclopropenone and triafulvene systems suggest some "cyclopropenium" character. Moreover, it is concluded[148] from the structural data, that the dipolar contribution is sensitively enhanced by weak intermolecular interaction such as hydrogen-bonding (as indicated by the bond-length variation when going from diphenyl cyclopropenone to its hydrate)[160].

From the evaluated structural parameters, CNDO/2 calculations were made giving information on charge separation in *11* and *182*[135]. Charge separation (expressed in q_0 values ≈ charge on oxygen) was found to be reasonably consistent within saturated ketones (acetone: –0.263, cyclopropanone: –0.249) and cyclopropenones (cyclopropenone: –0.365, dimethyl cyclopropenone: –0.371, diphenyl cyclopropenone: –0.386) and only slightly altered by C-O distance and C^1-C^3-C^2 angle variation. The significant increase of charge separation in cyclopropenones [*e.g.* q_0 (cyclopropenone) > q_0 (cyclopropanone)!] is apparently due to three-ring unsaturation. This means, however, that in contrast to Tobey's hypothesis[58] (p. 31) the charge magnitude cannot be solely a function of the carbonyl group.

The charge distribution in *11* and *182* was predicted by these calculations to be different despite the similiar electron-withdrawing capabilities of O (–0.387) and $C(CN)_2$ (–0.366) attributing a higher (+)-charge to the centers C^1/C^2 and the phenyl groups in the triafulvene than in the cyclopropenone. This is confirmed by ^{19}F-NMR measurements of the *p*-fluoro-phenyl substituted species (see p. 48) and by the reactivity of the triafulvene *182* toward nucleophiles (see p. 90), which attack exclusively at ring carbons C^1/C^2 and not likewise at C^3 as observed for the cyclopropenone *11*.

It was presumed by Ammon[135], that induction of larger amounts of charge separation than the ~40% values in *11/182/185* should be possible by introducing exocyclic residues derived from H_2X compounds of higher acidity than H_2O, $H_2C(CN)_2$, and tetrachlorocyclopentadiene. In view of this estimate of relative (–)-charge stabilization, triafulvenes *186/187* might be of particular interest.

CH₂(CN)₂ : pKₐ=11.2	CH₂(COCH₃)₂ : pKₐ=9.0	Dimedon : pKₐ=5.2
64	*186*	*187*

c) Studies Concerning the Electronic Structure of Cyclopropenone

According to the HMO treatment of cyclopropenone[58] and methylene cyclopropene[161] the following MO schemes have been derived together with bond orders, free valencies, and total π-electron energy.

MO scheme of cyclopropenone

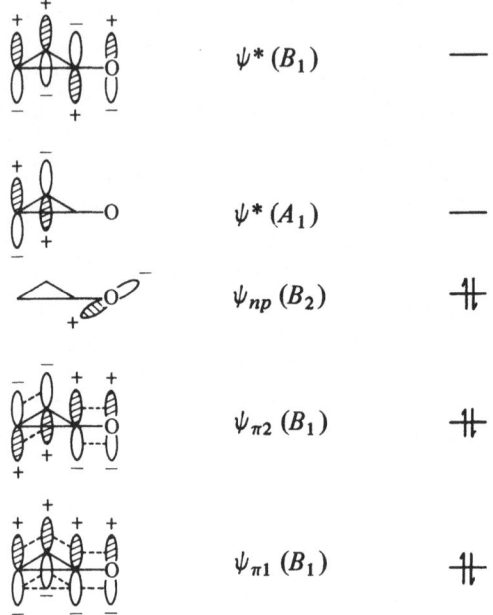

$\psi^* (B_1)$ —

$\psi^* (A_1)$ —

$\psi_{np} (B_2)$ ⇅

$\psi_{\pi 2} (B_1)$ ⇅

$\psi_{\pi 1} (B_1)$ ⇅

For comparison: MO scheme of the carbonyl group (C=O)

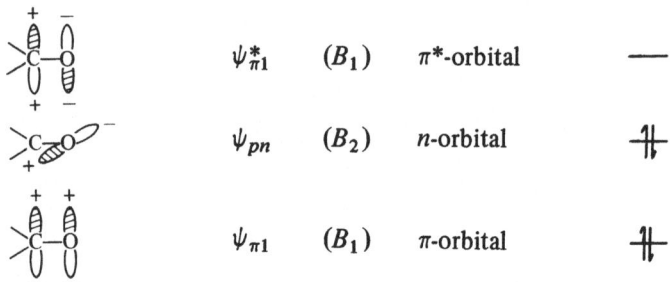

$\psi_{\pi 1}^*$	(B_1)	π^*-orbital	—
ψ_{pn}	(B_2)	n-orbital	⇅
$\psi_{\pi 1}$	(B_1)	π-orbital	⇅

Clark and Lilley[134] performed an *ab initio* LCAO-MO-SCF calculation on cyclopropenone regarding its geometry as that of cyclopropene with a C=O bond length of 1.21 Å as in formaldehyde. From the gross electron populations, a small σ-electron and an overall π-electron drift from the ring carbons to oxygen was deduced, the magnitude of which (−0.382) was close to the q_0 value of −0.365 for cyclopropenone and −0.385 for diphenyl cyclopropenone found by later calculations[135] based on CNDO/2.

MO scheme of methylene cyclopropene

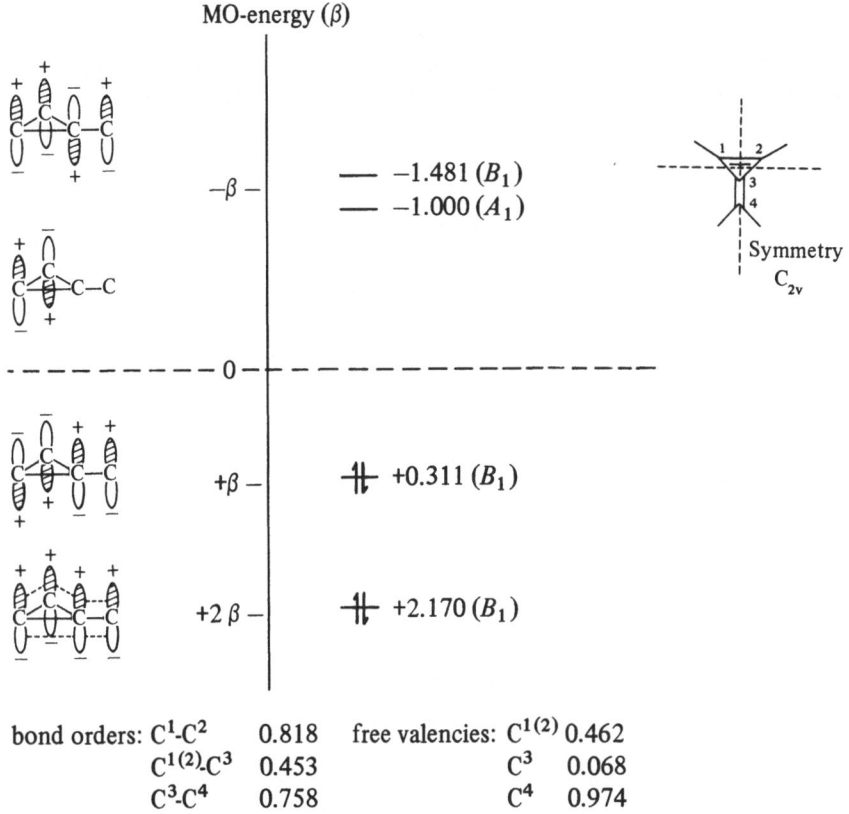

MO-energy (β)

$-\beta$ —

—— $-1.481\ (B_1)$
—— $-1.000\ (A_1)$

Symmetry
C_{2v}

0 —

$+\beta$ — ⵑ $+0.311\ (B_1)$

$+2\beta$ — ⵑ $+2.170\ (B_1)$

bond orders: $C^1\text{-}C^2$ 0.818   free valencies: $C^{1(2)}$ 0.462
            $C^{1(2)}\text{-}C^3$ 0.453 C^3 0.068
            $C^3\text{-}C^4$ 0.758 C^4 0.974

total π-electron energy: 4.962 β

From the electronic populations on the vinylic hydrogens, the acidity of vinylic C–H was estimated to be higher in cyclopropenone than in cyclopropene (0.684 e/ 0.776 e). This agrees with kinetic measurements of the H-D-exchange at n-propyl cyclopropenone[23] which showed an acidity of the vinylic C–H even higher than that of the acetylenic C–H in the reference compound propargyl alcohol.

The existence of significant delocalization in the π-system is indicated by the large localization energy of 341.6 kcal/mole for a lone-pair in a $2\,p_z$ orbital in the oxygen atom[138b]. Applying Koopman's theorem, the first two ionization potentials were calculated to be 9.54 eV (removal of a σ-electron) and 11.26 eV (removal of a π-electron). For cyclopropene, the assignment of the first two ionization steps is reversed giving 9.73 eV (π) and 11.42 eV (σ).

Similiar results on electron distribution of cyclopropenone, cyclopropenone imine and methylene cyclopropene were obtained in an *ab initio* calculation (SCF-MO) by Skancke[137a].

Robin *et al.*[162] investigated the photoelectron spectrum of unsubstituted cyclo-propenone and interpreted its results with the aid of Gaussian-type orbital calcula-tions of double-zeta quality for the electronic ground state using the experimentally established[133] geometry of cyclopropenone.

The spectra and calculations all led to the conclusion that there is an usually large interaction between both the π and lone-pair orbitals in the carbonyl portion of the molecule with the π and σ orbitals of the olefin portion. The first ionization potential (9.57 eV) involves ionization of an electron from the oxygen lone pair, whereas the second (11.19 eV) involves ionization of an electron from the olefin π-bond. The most vertical ionization is from the $7\,a_1$ MO (16.11 eV), the second lone-pair orbital on oxygen.

1st MO (symmetry $4\,b_2$): potential found 9.57 eV, calculated 9.61 eV.
2nd MO (symmetry $2\,b_1$): potential found 11.19 eV, calculated 10.40 eV.

The calculated MO's and the photoelectron spectrum confirm that the upper MO's in cyclopropenone are not at all localized.

(1) The "lone pair" (*n*-pair) at oxygen is intimately involved in the electron-bonding scheme of cyclopropenone: what is nominally the oxygen lone pair has only 54% of its density on oxygen, the remainder appearing in the σ-framework as C–C or C–H antibonding combinations.

(2) The $2\,b_1$ orbital has a large contribution from the C=C π-orbital (54%) with the remaining 46% being located totally on oxygen. Thus this orbital can be de-scribed as an almost equal mixture of the C=C and C=O π-bonds.

(3) The calculations showed that the C=C π-MO ionization in cyclopropene (9.86 eV) correlates most closely with the second ionization level in cyclopropenone and not with the first band. It is stressed that the noticably high dipole moment of cyclopropenone represents a manifestation of the extreme MO delocalization.

Information about inductive and conjugative interactions between C=C and C=O moieties has been derived from photoelectron spectra of di-tert. butyl cyclo-propenone and the corresponding cyclopropanone[163a] and correlated with MINDO/2 calculations.

Again, the lone pair at carbonyl oxygen is markedly delocalized. The carbonyl group exerts: (1) a stabilizing inductive effect, (2) a stabilizing conjugative effect [through bonding interaction between π (C=C) and π^* (C=O) MO's] on the C=C π MO. It is concluded that the cyclopropenone system bears some resemblance to the aromatic cyclopropenium system.

Photoelectron spectra of bis(dimethylamino) and bis(diisopropylamino) cyclo-propenone and bis(dimethylamino) cyclopropene thione have been measured and correlated with EHMO calculations[163b].

3. IR-Spectroscopic Investigations of Cyclopropenones and Triafulvenes

a) IR Spectroscopy of Cyclopropenones

Two very strong bands appearing in the region of 1830–1870 cm^{-1} and 1600–1660 cm^{-1} in the IR spectra are characteristic of cyclopropenones. The parent

	Observed	Calculated	
		ab initio	MINDO/2
n-MO	9.57	9.61	9.61 eV
π-(2 b_1)-MO	11.19	10.40	10.64 eV

	Observed	Calculated MINDO/2
n-MO	8.23	8.64 eV
π-(2 b_1)-MO	9.61	9.56 eV
Assignment from		n-MO at 8.45 eV
+ = C(CH$_3$)$_3$		

molecule is an exception showing high-energy transitions at 1833 and 1864 cm^{-1} and another strong band at 1480 cm^{-1}. [28]

The assignment of these absorptions has been controversial [2, 164, 165]: the low-energy transition near 1640 cm^{-1} has been assigned either to the C=O stretching mode or at least considered to have predominatly C=O character on the basis of solvent shifts.

According to investigations by Krebs [143, 166-168] the C=C and C=O double bonds are strongly coupled over the single bonds of the very rigid cyclopropenone system as indicated by the following findings:

(1) ^{18}O substitution in dimethyl and diphenyl cyclopropenone affects both strong bands by significant shifts to lower wavenumbers, but the shifts are smaller (10–20 cm^{-1}) than in ordinary ketones ($\sim$30 cm^{-1}). This can only be explained by strong mixing of several modes.

(2) In the Raman spectra of cyclopropenones the absorption near 1640 cm^{-1} displays by far the highest intensity, which should not be expected for a pure C=O stretching mode.

(3) The positions of both absorptions in methyl and n-propyl cyclopropenone (1838/1605 cm^{-1}, 1835/1600 cm^{-1}) are displaced to lower wavenumbers compared to the symmetrically substituted species (dimethyl cyclopropenone: 1848, 1866/1657 cm^{-1}; di-n-propyl cyclopropenone: 1840/1630 cm^{-1}). If the 1640 cm^{-1} band were a pure C=O stretching vibration, the observed large effect of three-ring substitution could not be explained, since no change in the relative mass and geometry of participating atoms takes place.

(4) A normal coordinate analysis has been carried out for dimethyl and diphenyl cyclopropenone, which gave a potential energy distribution of vibrations as follows: 1850 cm^{-1} band: 40–50% on C=O, 10–20% on C=C, and 30–35% on C–C (ring); 1640 cm^{-1} band: 20–30% on C=O, 55–60% on C=C, and about 0% on C–C (ring). A third band at 880 cm^{-1} in the Raman spectrum is distributed over the same bonds:

15–20% on C=O, about 15% on C=C, and 55–65% on C–C (ring). The calculated eigenvectors show that in the 1850 cm^{-1} band the phase of the C=O group is opposite to that of the bonds in the ring, whereas in the 1640 cm^{-1} band the C=O and C=C groups are vibrating in phase.

Recently, a normal coordinate analysis has been done for cyclopropenone itself and its dideuterated derivative[169] assigning the unusually low band at 1480 cm^{-1} to "(C=C)" frequency with strong vibrational coupling. The force constants derived were believed to be in accordance with a substantial contribution of a zwitterionic form to the electronic ground state of cyclopropenone. As already concluded by Krebs[167] "the 1850 cm^{-1} band can be assigned to an out-of-phase stretching vibration of the two double bonds with a predominance of the C=O coordinate and the 1640 cm^{-1} band to the corresponding in-phase vibration with a predominance of the C=C coordinate. The band at 880 cm^{-1} is mainly a symmetric stretching vibration of the C–C bonds in the ring. All three vibrations must be considered as typical features of the cyclopropenone system".

b) IR Spectroscopy of Triafulvenes

The two bands which appear in the region of 1810–1880 cm^{-1} and 1510–1550 cm^{-1} can be regarded as characteristic of the IR spectra of the triafulvene system, as documented by the following examples of alkyl-substituted triafulvenes[79, 170]. Monophenylsubstituted 4,4-diacyl triafulvenes[55] are exceptions in that the high-energy transition is displaced to 1770–1780 cm^{-1}.

CH$_3$...CH$_3$ / NC...CN	n–C$_3$H$_7$...n–C$_3$H$_7$ / NC...CN	For comparison:	Ph...Ph / NC...CN
1870	1879		1890
1510	1579[74]		1552[67]

CH$_3$...CH$_3$ / NC...CO$_2$CH$_3$	CH$_3$...CH$_3$ / NC...COPh
1870	1865
1515	1610 (C=O)
1680 (C=O)	
188	*189*

It seems to be likely that by analogy with cyclopropenones the observed absorptions originate from strong coupling of the endo- and semicyclic C=C bonds; however a detailed IR and Raman analysis of triafulvenes has not yet been performed.

It should be noted that functional groups, *e.g.* C=O at the exocyclic carbon (C^4) of triafulvenes often show characteristic shifts to lower frequency compared to the corresponding bands in the "non-cyclopropenylated" systems. Thus, in tria-

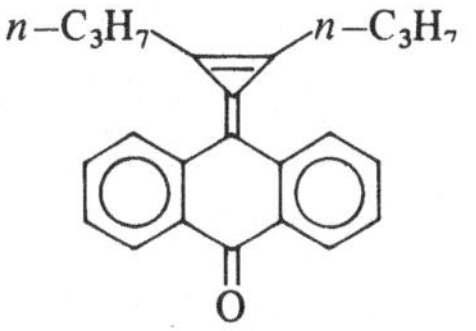

1820	1858	1859
1630	1527	1553
190		

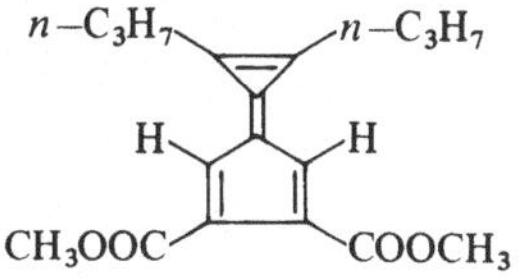

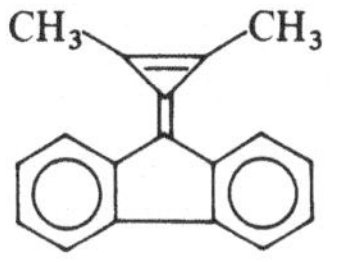

1850	1830	1770
1620/1660(C=O)	1620/1660(C=O)	1605/1640(C=O)
	191	

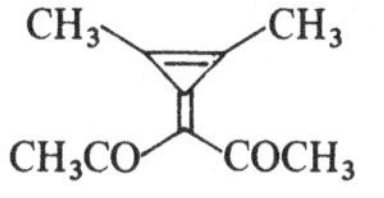

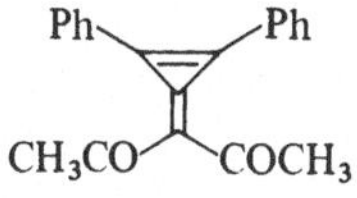

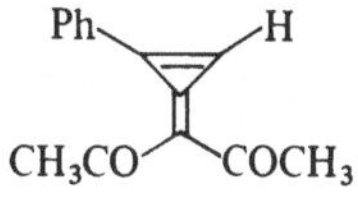

fulvenes *190* and *188* carbonyl absorption is found at 1630 and 1680 cm^{-1}, respectively; however, C=O groups in anthrone (1670 cm^{-1}) and methyl cyanoacetate (1770 cm^{-1}) absorb at markedly higher frequency. This may point to an enhanced single-bond character for the exocyclic C=O group in the triafulvenes, which supports the idea of "cyclopropenium enolate" participation in the triafulvene resonance hybrid, as shown in *192a/192b*:

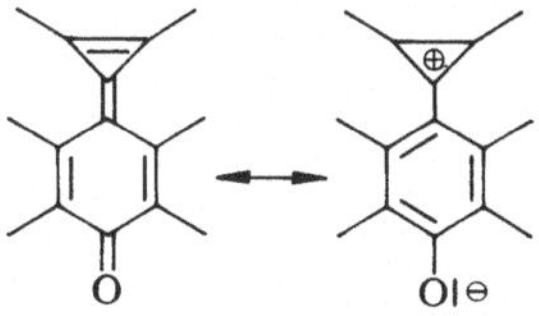

192a 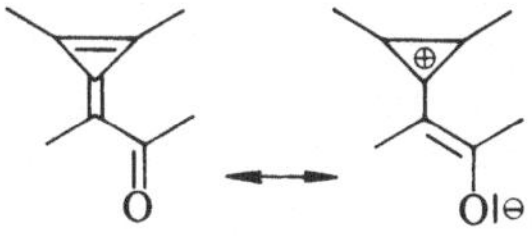   *192b*

4. UV Spectroscopy of Cyclopropenones and Triafulvenes

a) Cyclopropenones

In the parent molecule, an absorption maximum at 276 nm (log ϵ = 1.49) and end absorption below 190 nm were found and were assigned to $n \rightarrow \pi^*$ and $\pi \rightarrow \pi^*$ transitions, respectively[28].

In dialkyl cyclopropenones, the $n \rightarrow \pi^*$ transition was found in the region of ~250 nm and again there was strong end absorption resulting from $\pi \rightarrow \pi^*$ transition. The relatively high transition energies are consistent with MO calculations, which predict a relatively high energy for the corresponding antibonding MO in cyclopropenone.

41

The $n \rightarrow \pi*$ transition appears at a considerably lower wave-length than in other α,β-unsaturated ketones (310–320 nm) and cyclopropanone (310 nm) and shows a marked negative solvatochromy (hypsochromic shift of 41/35 nm for cyclohexane-water for the systems in Table 8). Since the direction of the solvent shift points to a stronger stabilization of the ground state when going to more polar media capable of hydrogen-bonding, the observed effect is believed[143] to support further evidence for dipolar contributions to the cyclopropenone ground state.

Table 8. $n \rightarrow \pi*$ transition in dimethyl and cyclohepteno cyclopropenone and its solvent dependence (from Ref.[143], extinctions in log ϵ)

Solvent	Dimethyl Cyclopropenone	Cyclohepteno cyclopropenone
Cyclohexane	272 (1.63)	267 (1.90)
Dichloromethane	257 (1.36)	254 (1.89)
Ethanol (95%)	243 (1.69)	243 (1.92)
Water	231 (1.78)	232 (1.92)

UV comparison of some diaryl cyclopropenones

R = phenyl	297	285	220 nm	
R = p-toly	308	297	240	227 nm
R = p-anisyl	342	315	256	233 nm
		296		
R = p-chloro phenyl	310	300	236	227 nm
Cis-stilbene		277	224	202 nm
Diphenyl cyclo-propenium cation[171]	306	293 nm		

In diaryl cyclopropenones, the UV data are not very characteristic and resemble those of diaryl cyclopropenones and stilbenes as well as diaryl cyclopropenium cations, as shown in Table 8. Recently a new long-wave absorption band was found[164] in diphenyl cyclopropenone [362 nm, log ϵ = 3.06 (cyclohexane)] which was assigned tentatively to an intramolecular charge-transfer.

b) Triafulvenes

The UV spectra of most of the triafulvenes hitherto known are rather complex in structure due to effects of either phenyl substitution or of conjugated chromophores at the exocyclic carbon, which obscure information on the cross-conjugated system itself.

The UV data of the dialkylsubstituted species 182 and 193–195 can be regarded to be more specific for the triafulvene system.

$n-C_3H_7$ ⟩═⟨ $n-C_3H_7$ / NC–CN

193[74]

cyclohexane
246 nm (4.30)

methanol
245 nm (4.33)

CH_3 ⟩═⟨ CH_3 / CH_3CO–$COCH_3$

194[170]

dichloromethane
274 nm (4.09)

acetonitrile
271 nm (4.13)

methanol
271 nm (4.12)

CH_3 ⟩═⟨ CH_3 / O═ ═O / H_2 H_2 / CH_3 CH_3

195[59]

dichloromethane
241, 285 nm (4.04, 4.20)

acetonitrile
242, 283 nm (4.18, 4.30)

methanol
246, 282 nm (4.24, 2.36)

CH_3 ⟩═⟨ CH_3 / NC–CN

182[79]

dichloromethane
246 nm (4.27)

acetonitrile
243 nm (4.29)

methanol
243 nm (4.24)

Ph ⟩═⟨ Ph / CH_3CO–$COCH_3$

180[119]

dichloromethane
308 nm (4.43)

acetonitrile
306 nm (4.47)

methanol
307 nm (4.44)

If the bathochromic effect of geminal dicyano substitution is assumed to be in the same range as found for ethylenic reference compounds[79], from the positions of maxima in *182* and *193* the absorption of the parent triafulvene can be estimated to appear in the region of 205–215 nm. This is in qualitative agreement with MO prediction[172] that the $\pi \to \pi^*$ transition for methylene cyclopropene will fall near 200 nm.

Interestingly, the UV maxima of the above dialkyl triafulvenes are only slightly influenced by the polarity and hydrogen-bonding capability of the solvent. In the cyclic triafulvene *195*, countercurrent solvent shifts for the two main absorptions are observed.

In contrast, the diphenyl analogue of *182/192* and other triafulvenes[5] showed marked solvent dependence of UV absorption, whereas the diphenyl analogue of *194* does not (see above). The origin of this complex behavior is still an open question.

Some aspects of the UV spectroscopy of calicenes and quinocyclopropenes deserve attention. Thus, the UV data of the tetrachloro compound *184* and the thioketal *196* suggest[93] that the absorption maximum of the parent pentatriafulvene (*197*) should lie close to 300 nm:

λ_{max}=314nm

184

λ_{max}=310nm

196

197

If this estimate is correct, the transition energy (4.1 eV) is significantly higher for pentatriafulvene than for the lowest energy singlet transition of fulvene (3.4 eV). Interestingly this order of transition energy is predicted by PPP-SCF calculations[93].

Föhlisch[99] reported a remarkable dependence of the electron spectra of quino-cyclopropenes on their structure. As shown in Table 9, the merocyanine-like quino-cyclopropenes show positive solvatochromy when they contain an anthraquinonoid chromophore (198), but negative solvatochromy when they contain a benzoquino-noid system (199). This can be interpreted in terms of a markedly increased partici-pation of dipolar resonance forms in the ground state of the benzoquinonoid 199 compared to the anthraquinonoid 198. From the dipole moment of 198 (9.4 D^{99}) the dipolar contribution was estimated to be in the range of ~23%.

R=n–C$_3$H$_7$: 190

R=Ph   : 198

199

These findings point to a general problem of the triafulvene system, namely that its degree of "polarity" is considerably affected by C^4-substituent properties as in the case of quinocyclopropenes 198/199 by the different aromatization ten-dency of the quinonoid system.

Table 9. Solvent dependence in quinocyclopropenes 190/198/199 (from Ref.[99])

Solvent	199	198	189
Cyclohexane	410, 388 nm	426 nm	420, 397 nm
Benzene	410, 389 nm	440 nm	435, 411 nm
Dichloromethane	406, 385 nm	448 nm	443, 421 nm
Acetonitrile	402, 382 nm	447 nm	442, 425 nm

5. NMR Spectroscopy of Cyclopropenones and Triafulvenes

a) ^{1}H-NMR Spectroscopy

Information specific to the electronic and bonding properties of the cyclopropenone and triafulvene system have been derived mainly from species bearing alkyl groups or protons on the three-membered ring. Since phenyl groups (giving an unspecific A$_2$B$_2$X splitting pattern of aromatic protons) and substituents at exocyclic carbon in triafulvenes have been found to be not particularly informative[119], discussion is restricted to the alkyl- or H-substituted cases, for which ^{1}H-NMR data are presented in Table 10.

Table 10. Vinylic C–H proton and α–C–H chemical shift in monosubstituted cyclopropenones and triafulvenes

	R = H	$0.92\ \tau\ (CH_3NO_2)$[175]
		--
	R = CH$_3$	$1.34\ \tau\ (CCl_4)$[23]
		$7.60\ \tau\ (CH_3)$
	R = n-C$_3$H$_7$	$1.32\ \tau\ (CCl_4)$[23]
		$7.28\ \tau\ (CH_2(\alpha))$
	R = n-C$_5$H$_{11}$	$1.53\ \tau\ (CDCl_3)$[48]
		$7.30\ \tau\ (CH_2(\alpha))$
		$1.32\ \tau\ (CCl_4/CF_3COOH)$
		$7.17\ \tau$
	R = C$_6$H$_5$	$1.55\ \tau\ (CDCl_3)$[55]
		--

R = (steroid, OAc / AcO) $1.68\ \tau\ (CDCl_3)$[26]   --

R = (steroid, HO / AcO) $1.57\ \tau\ (CDCl_3)$[26]   --

R = C$_6$H$_5$ $1.15\ \tau\ (CDCl_3)$[55]   --

α–C–H chemical shift in dialkyl cyclopropenones and dialkyl triafulvenes

R = CH$_3$	$7.75\ \tau\ (CCl_4)$[23]	R = n-C$_4$H$_9$ $7.60\ \tau\ (CCl_4)$[43]
R = n-C$_3$H$_7$	$7.43\ \tau\ (CCl_4)$[43]	R = cyclo-hepteno $7.45\ \tau\ (CDCl_3)$[43]
R = CH$_3$, R' = CN	$7.44\ \tau\ (CDCl_3,\ CH_3)$[79]	
R = n-C$_3$H$_7$, R' = CN	$7.13\ \tau\ (CHCl_3,\ CH_2(\alpha))$[74]	
R = CH$_3$, R' = COCH$_3$	$7.26\ \tau\ (CDCl_3,\ CH_3)$[170]	

As Table 10 shows, vinylic three-ring protons in cyclopropenone itself and in monosubstituted cyclopropenones appear at remarkably low field in the region of 0.9–1.6 τ. Comparison with protons at the double bond of covalent cyclopropenes (e.g. *200*: 3.34 τ[174]) and cyclopropenium cations (*210*: –0.35 τ[146]; *202*: –1.20 τ[173]) shows the cyclopropenone system to exhibit a far stronger deshielding effect (~2 ppm) than the cyclopropene moiety, but weaker than a cyclopropenium ring system.

The ^{13}C–H coupling constants of methyl (213 Hz[23]) and phenyl (216 Hz[55]) cyclopropenone are in the order of those obtained for cyclopropene vinylic protons (*200/201*: 218 Hz/221 Hz[174]) and reflect an *s*-contribution of more than 40% in the carbon hybrid orbital of the vinyl C–H bond.

In the monophenyl triafulvene *190* ($J_{13_{C-H}}$ = 232 Hz[55]) *s*-character is higher and corresponds to the value of 230 Hz found for cyclopropenone.

Intriguingly ^{13}C–H coupling of cyclopropenes *203/204* substituted at C^3 by strongly electron-withdrawing groups comes even closer to the value (265 Hz[173])

CH_3, (H) ⟶ 3.34 τ

J_{13C-H} = 218 Hz

CH_3 CH_3

200

H, (H) ⟶ 2.68 τ

J_{13C-H} = 221 Hz

CH_3 CH_3

201

H, (H) ⟶ -1.20 τ (CH_3NO_2)

J_{13C-H} = 265 Hz

H

202

H, H

X X

203 : X=Cl, J_{13C-H} = 239 Hz[175)]

204 : X=CN, J_{13C-H} = 255 Hz[176)]

observed for the delocalized 2π cyclopropenium cation. Since, however, diethyl cyclopropenium cation is reported[177a)] to have J_{13C-H} = 228 Hz, a relationship between J_{13C-H} and "cyclopropenium character" does not seem to be reasonable.

The chemical shift of methyl groups and the "internal" chemical shift (α-CH_2 versus β-CH_2 protons) of n-propyl groups has been used as a probe for the π-electron density of cationoid centers. Therefore, the ^{1}H-NMR spectroscopic behavior of methyl and n-propyl substituted cyclopropenones and triafulvenes is of particular interest (see Table 10) with reference to the cyclopropene carboxylic acids *205/206* (*205*: CH_3 7.95 τ[178)]; *206*: $\Delta CH_2(\alpha/\beta)$ = 0.80 ppm[43)]) and the methyl and n-propyl substituted cyclopropenium cations *207–210* (*207/208*: $CH_3 \sim 7.0 \tau$[177b, c)]; *209/210*: $\Delta CH_2(\alpha/\beta)$ = 1.27 – 1.30 ppm[146)]).

R, R

H COOH

205: R = CH_3

  α β
206: R = $CH_2CH_2CH_3$

R, R

R'

$\Delta CH_2(\alpha/\beta)$

R = CH_3		R = n-C_3H_7
207: R' = CH_3		*209*: R' = n-C_3H_7
208: R' = H		*210*: R' = H

As seen from Table 10, both the methyl resonance in dimethyl cyclopropenone (7.75 τ) and the separation of CH_2 units α and β to the three-ring in di-n-propyl cyclopropenone (0.85 ppm) compare well to corresponding values for the covalent cyclopropene derivatives, but differ strongly from those of the positively charged cyclopropenium species.

In monomethyl and mono n-propyl cyclopropenones, the down-field shifts of CH_3 and α-CH_2 groups as well as the CH_2 (α/β) separation (0.94 ppm) are somewhat larger than in the dialkyl case, but only to about 50% of the relative displacement of vinylic hydrogens (distance vinyl-H (*200*) versus vinyl-H (*210*) taken as a reference (3.70 ppm)).

The dimethyl substituted triafulvenes *211a–e* (Table 11) show methyl resonances considerably shifted downfield from dimethyl cyclopropenone. The same tendency is observed for the α-CH$_2$ protons in the di-n-propyl substituted triafulvene *192*[74)] along with an enlarged CH$_2(\alpha/\beta)$ separation value.

An interpretation of the rather complex chemical shift data can by no means be straight forward. The observed values reflect appreciable electron-deficiency at the three-ring carbons, in part some positive charge delocalization and a small ring current, which is evident at least for the H-substituted species. However, they should be influenced — to an extent difficult to estimate — by effects of the magnetic anisotropy of the carbonyl group in cyclopropenones and exocyclic C^3 substitutents (*e.g.*, containing CN and CO) in triafulvenes. Thus, since an even qualitative differentiation between the various effects cannot be reached, a simple interpolation[179)] to the amount of "cyclopropenium character" merely from chemical shift data is regarded to be unsafe[23)].

Likewise, a consistent relationship between relative displacement of three-ring substituents and the ability of the exocyclic part X in triafulvenes and cyclopropenones to stabilize the (−) charge (expressed by the pK$_a$ of the H$_2$X compounds derived from it[135)]) is not fulfilled, as demonstrated by Table 11. The practically identical chemical shifts in *211c/d/e*, differing strongly in acidity of the corresponding H$_2$X compounds, may be due to superimposition of the anisotropy of exocyclic substituents to ring current and charge separation involved.

Table 11. Comparison of CH$_3$ chemical shift and difference to reference *200* of several dimethyl-substituted triafulvenes to pK$_a$ of the corresponding H$_2$X compounds

		CH$_3$ (τ)	$\Delta\tau$	pK$_a$[180)]
	X = O[43)]	7.75	0.22	15.7
a)	X = C(CN)$_2$[79)]	7.44	0.53	11.2
b)	X = C(CN)(COOCH$_3$)[79)]	7.42 / 7.39	0.57 (aver.)	9
c)	X = C(COCH$_3$)$_2$[170)]	7.26	0.71	9.0
d)	X = C(CN)(NO$_2$)[55)]	7.24 / 7.34	0.68 (aver.)	~7
e)	X = (dimedone-type)[59)]	7.29	0.68	5.2

b) ^{13}C-NMR Spectroscopy

^{13}C-NMR data are available only for cyclopropenone itself[133)] and several dialkyl-substituted cyclopropenones[47)] as given in Table 12.

Table 12. ^{13}C-NMR of cyclopropenones

	C^1(2)	C^3	(δ, from TMS reference, CDCl$_3$)
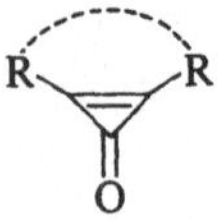 R = H	158.3	155.1	
R = C(CH$_3$)$_3$	164.8	159.5	
R = –(CH$_2$)$_5$–	164.2	154.6	
R = –C(CH$_3$)–(CH$_2$)$_2$–C(CH$_3$)– (with CH$_3$ groups)	169.0	146.7	

The marked shifts observed when going from the seven-ring cyclopropenone *39* to the six-ring cyclopropenone *40* have been interpreted as an indication of a change in electron distribution: for *40* a stronger contribution of resonance structures *b* might lead to relief of strain and thus stabilize the fused-ring system.

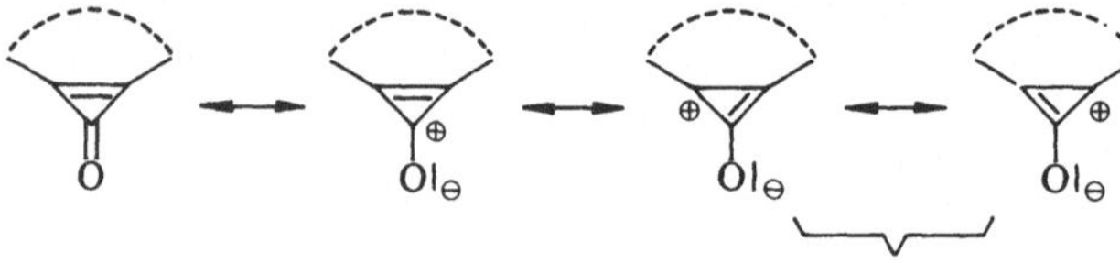

Although this interpretation is regarded as tentative[47], it is assisted by the unique reactivity of the strained cyclopropenone *40* toward OH-ions (see p. 67).

c) ^{19}F-NMR Spectroscopy

^{19}F-shifts in *p*-substituted fluorobenzenes are supposed to provide a measure of the σ- and π-electron withdrawing properties of groups para to the F atom[181]. Bis(*p*-fluorophenyl)cyclopropenone and 1,2-bis(*p*-fluorophenyl)-4,4-dicyano triafulvene (*212/213*) were found to exhibit ^{19}F resonances at 61.5 and 65.5 ppm (downfield from C$_6$F$_6$)[182].

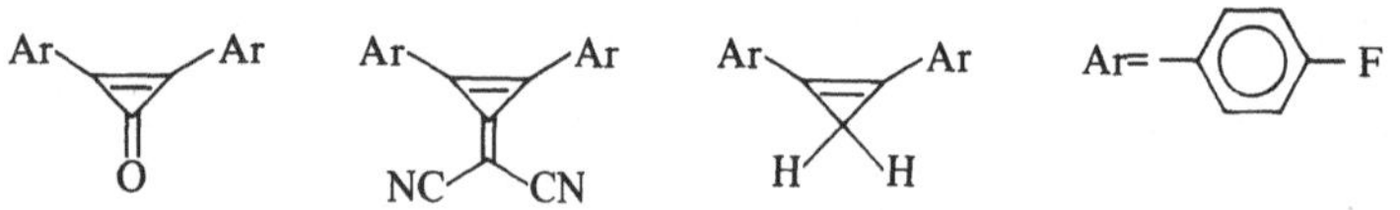

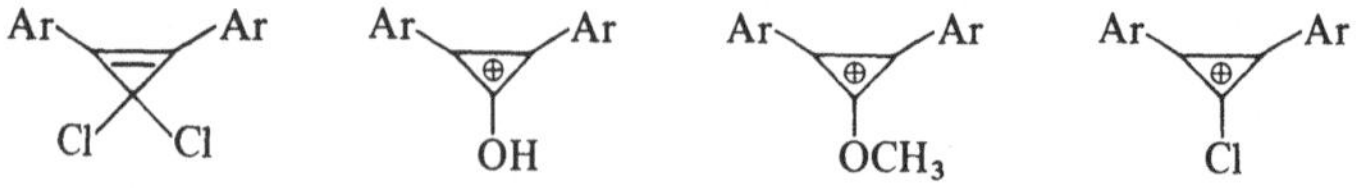

The observed difference (*212* vs. *213*) of 4.0 ppm is in qualitative agreement with the prediction of CNDO/2 calculations[135] (see p. 35) indicating a greater electron deficiency at C^1/C^2 and in the phenyl rings of the triafulvene than in the cyclopropenone.

Tobey[58] has reported further ^{19}F-NMR studies comparing bis(*p*-fluorophenyl)-cyclopropenone (*212*) to its dichloride *214* and the cations *215–217*. Although the cyclopropene *218* should be a less ambigous reference than the dichloride *214*, it can be concluded that the deshielding effect of the cyclopropenone system is related closer to the covalent cyclopropene derivative than to the cationic species. This is in qualitative accordance with the findings in Chapter 5 (a).

d) Dynamic NMR Properties of Triafulvenes

As a result of the small, but apparent single bond character of the triafulvene C^3/C^4 bond and the good stabilization of the transition state of the rotation established earlier, rotation around this bond should be lower in energy in comparison to simple ethylene derivatives[183]. In fact, ^{1}H-NMR spectra of several types of asymmetrically substituted triafulvenes *219–224* proved to be temperature-dependent and showed reversible coalescence phenomena at definite temperatures diagnostic for internal rotation processes. These were characterized by the free enthalpy of activation $\Delta G_c^{\#}$ at the coalescence point of appropriate substituent signals[61].

Ar = *p*-tolyl, *p*-anisyl, *p*-(tert. butyl)phenyl
R = CH_3, $C(CH_3)_3$, α-naphthyl
R′ = alkyl, aryl, NH–Ph, O-alkyl
R″ = CN, NO_2, acyl, aroyl

Triafulvenes of type *224* capable of cis-trans "rotational" isomerism in most cases occur as equilibrium mixtures of structures *224*$_A$ and *224*$_B$; their configurational and conformational assignment was accomplished unambiguously, but separation into their stereoisomeric components was not possible[55]. Triafulvenes *225/226*, however, crystallized as one definite configurational isomer[55, 63] which equilibrated in solution with its "rotamer" thus allowing a kinetic determination of $\Delta G^{\#}$ more reliable than that from coalescence methods:

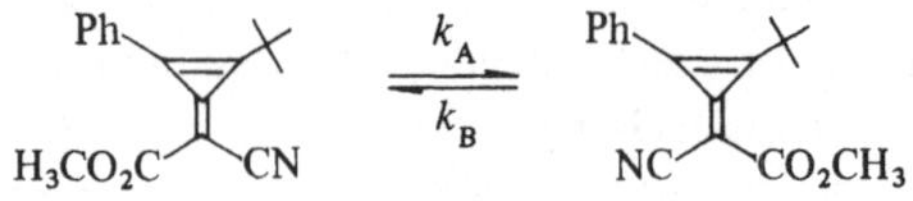

225

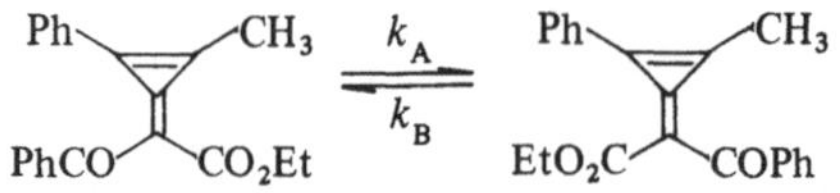

226

k_A $= 0.925 \cdot 10^{-4}$ (sec^{-1}); $\Delta G_A = 28.8$ kcal/mole

k_B $= 0.812 \cdot 10^{-4}$ (sec^{-1}); $\Delta G_B = 28.9$ kcal/mole
$t_{1/2}$ $= 66.5$ min, temp. 100 °C

k_A $= 0.442 \cdot 10^{-3}$ (sec^{-1}); $\Delta G_A = 22.4$ kcal/mole

k_B $= 0.694 \cdot 10^{-3}$ (sec^{-1}); $\Delta G_B = 22.1$ kcal/mole
$t_{1/2}$ $= 10.2$ min, temp. 50 °C

As indicated by the observed $\Delta G^{\#}$ values, internal rotation of triafulvenes is influenced by electronic and steric parameters of substituents, the latter in some cases (*e.g. 222*) accounting for an enhanced ground state energy relative to the transition state[61].

In addition, a marked influence of solvent polarity on $\Delta G_c^{\#}$ was found to be exemplified for triafulvene *227*, which showed a decrease of $\Delta G_c^{\#}$ when measured in solvents with higher dielectric constants (Table 13).

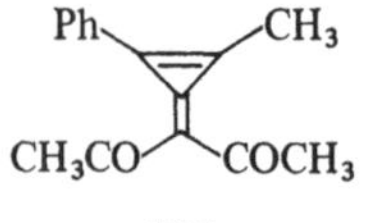

227

The apparent lowering of the rotational barrier in triafulvenes is open to interpretation either by substituent or solvent stabilization of ground-state polarity leading to a decrease of C^3/C^4 "double" bond character or by stabilization of a more polar – probably perpendicularly orientated[184] – transition state by substituent or solvent effects.

Table 13. Coalescence measurements of 1-methyl-2-phenyl-4,4-diacetyl triafulvene (227) in different solvents (from Ref.[61])

Solvent	DK	$\Delta\nu(COCH_3)^{(Hz)}$	T_c (°C) (coalescence temperature)	$\Delta G_c^{\#}$ ($\frac{kcal}{mole}$)
Formamide	109.5	7	73.5	18.5
Nitrobenzene	34.8	11.2	86.5	18.9
Benzonitrile	25.2	12.3	135	21.6
Benzylcyanide	18.7	18.5	–	–
Chlorobenzene	5.6	23	145	21.8
1-Chloro-naphthalene	5.04	35	180	23.1
Benzene	2.28	41	–	–

A similiar situation is found in the calicene series. An all-electron calculation on the energy barrier in unsubstituted calicene (197) was carried out[185] and implied a zwitterionic perpendicular structure in the resonance hybrid with appreciable hyperconjugative interaction. The obtained value of 26.8 kcal/mole was used for estimating the rotational barrier of 1,2,3,4-tetrachloro-5,6-dialkyl calicenes 228 to be in the range of ~16 – 19 kcal/mole by PPP-SCF calculations[184]. This lowering of rotational energy by inductive effects of substituents at the five-membered ring compares well with the $\Delta G_c^{\#}$ value of ≈19 kcal/mole found for the calicene 229[186], which is stabilized by the electron-withdrawing formyl residue in the 1-position.

197   228   R=alkyl $+$ =C(CH_3)_3

$\Delta G_c^{\ddagger}$ ≈19 kcal/mole

229   230

The di-tert. butyl substituted calicene 230 was calculated to possess considerable non-bonding (steric) interactions in the planar geometry[184]. The relief of strain when going to a perpendicular transition state is reflected by the low coalescence temperature of tert. butyl signals found on temperature-dependent ¹H-NMR spectroscopy[187].

51

Cyclopropenium-immonium cations *231* and *232* have been found to behave in analogy to triafulvenes according to investigations by Krebs[45].

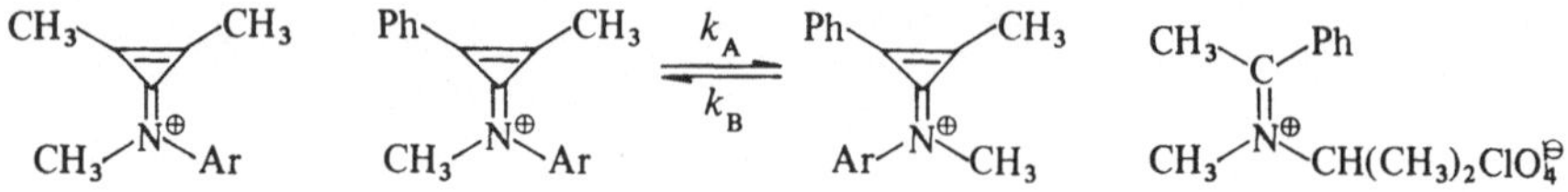

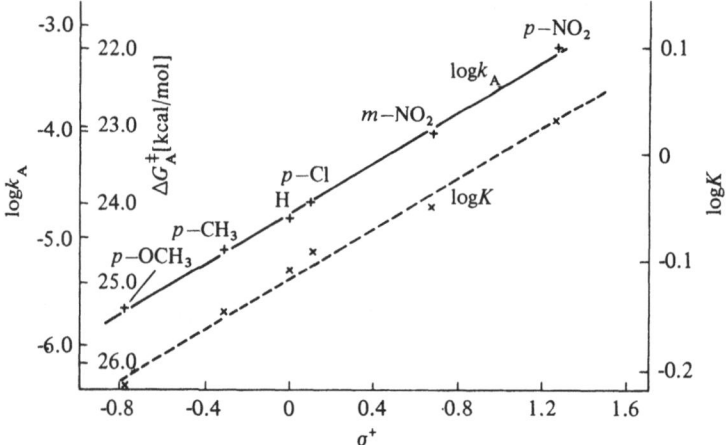

$$\Delta G_c^{\ddagger} \sim 23\text{-}25$$

231

$$\Delta G^{\ddagger} \sim 22\text{-}25$$

232

$$\Delta G^{\ddagger} \geq 30 \text{ kcal/mole}$$

233

Ar = *p*-X-phenyl; X = H, CH_3, OCH_3, NO_2, Cl

The $\Delta G^{\#}$ values for the rotation around the C^3–N bond were obtained by the coalescence method and kinetic measurements of the equilibration of isolated and configurationally established cis-trans isomers of type *232*. The barrier of rotation is considerably lower than in ordinary immonium cations, e.g. *233*; $\Delta G^{\#}$ is decreased by electron-withdrawing substituents at the nitrogen atom, whilst at the three-ring the opposite effect is observed.

The log k and $\Delta G^{\#}$ values of immonium cations of type *232* and the log K (equilibrium) values can be correlated (see diagram below) satisfactorily with the σ^+ constants of Brown.

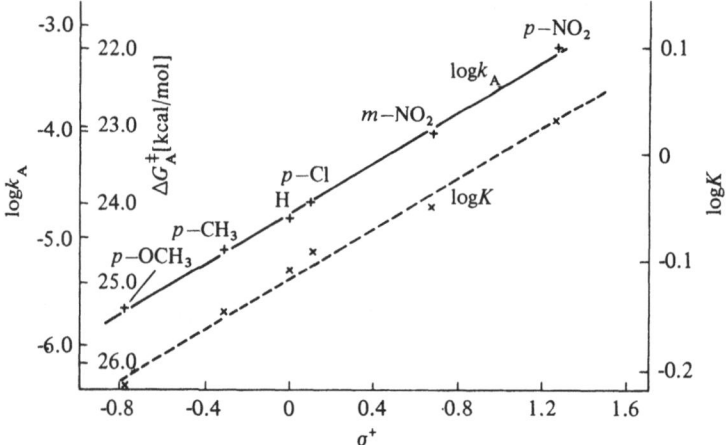

6. Mass Spectrometry of Cyclopropenones and Triafulvenes

Mass spectra for a number of cyclopropenones have been reported. Thus, for dimethyl[23], di-tert. butyl[10], di-cyclopropyl[57], di(*p*-anisyl)[55], monomethyl[23], and mono n-propyl[23] cyclopropenone the molecular ion has been found; however, for diphenyl[10], methyl phenyl[55], tert. butyl phenyl[63] and phenyl[54] cyclopropenone as well as for diphenylcyclopropene thione[188] the first fragment corresponds to initial loss of CO (CS) and is followed by the fragmentation of the acetylenic residue.

It is uncertain whether this primary process is thermally induced or electron-impact induced (see also p. 55).

For dichlorocyclopropenone[32] a fragmentation pattern corresponding to the loss of Cl, Cl_2, CO, COCl, and C_2ClO was observed, which does not seem to be characteristic for the cyclopropenone system.

Analogy to cyclopropenones was found in electron-impact studies of triafulvenes[55]. Generally fragment ions 234 derived from cycloreversion to alkynes were generated, except in the case of the methyl-substituted species, which stabilizes fragment 234 by loss of hydrogen forming phenyl cyclopropenium ion 235[189]:

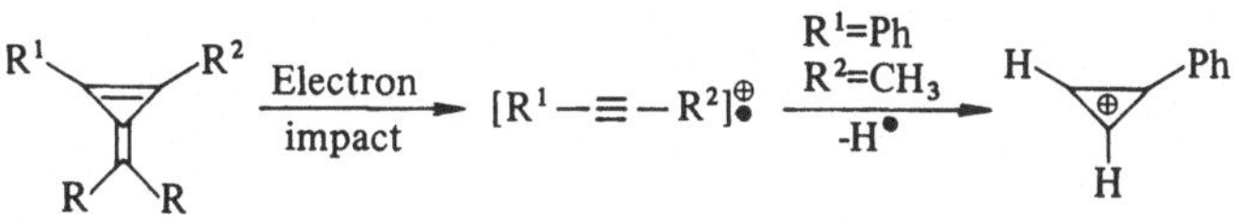

R¹   R²   Electron
  ⟩⟨   ────────→ [R¹—≡—R²]⊕•
 R   R   impact

R¹=Ph
R²=CH₃
─────→
-H•

234   235

However, mass spectrometry also revealed some unexpected properties of triafulvenes. Thus benzotriapentafulvalene 236 showed a fragmentation pattern including loss of 2 CO and 2 hydrogens, which was interpreted via intramolecular hydrogen transfer and ring-contracting decarbonylation to 237, ending with a fragment of the elusive benzotriafulvalene structure 238[80,190]. This mechanism is supported by experiments with the deuterated triafulvene 236.

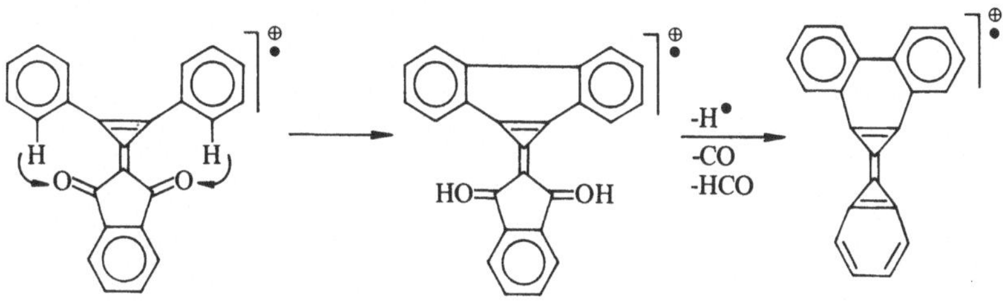

236   237   238

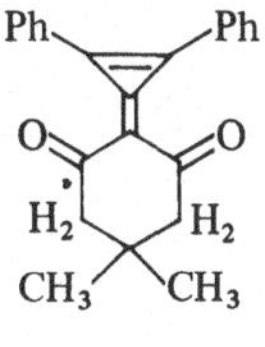

84

The same fragmentation type is observed for triafulvene 84 derived from dimedone[55].

A different fragmentation was found for 4-acyl triafulvenes 239[55]. As depicted below, primary loss of hydrogen (M^+-1) is followed by elimination of carbon monoxide (M^+-29), which is only reasonable if the phenyl residue in the (M^+-29)-fragment is attached to the triafulvene system by a second binding site as considered in 240. Further evidence for a highly conjugated "structure" (240) of this fragment is brought about by the corresponding doubly charged ion appearing in relatively high intensity. Again, the mechanism is confirmed by experiments with deuterated species.

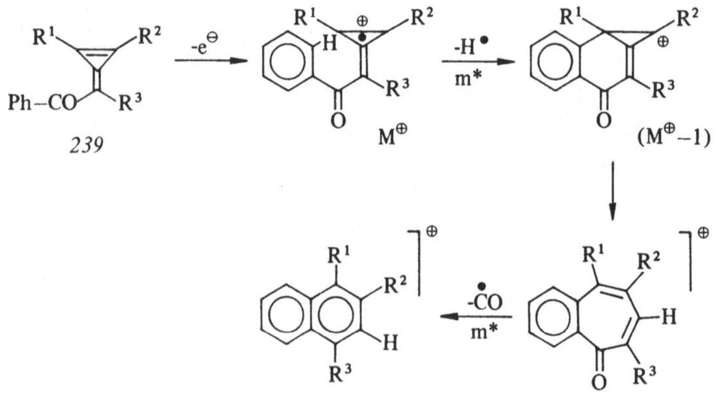

239

M⊕

(M⊕−1)

240

With 4,4-diacyl triafulvenes two principal fragmentation pathways have been observed[55]. In 4-aroyl-4-acetyl triafulvenes *241* the molecular ion is followed by a fragment ion of probable "structure" *242* arising from primary loss of $(C_7H_6R)^{\cdot}$, which surprisingly has incorporated a CH_2 unit from the acetyl group and the exocyclic aryl residue. It is not unlikely that the (C_7H_6R)-residue corresponds to a substituted tropyl radical due to its well-known formation from electron-impact of benzylic precursors.

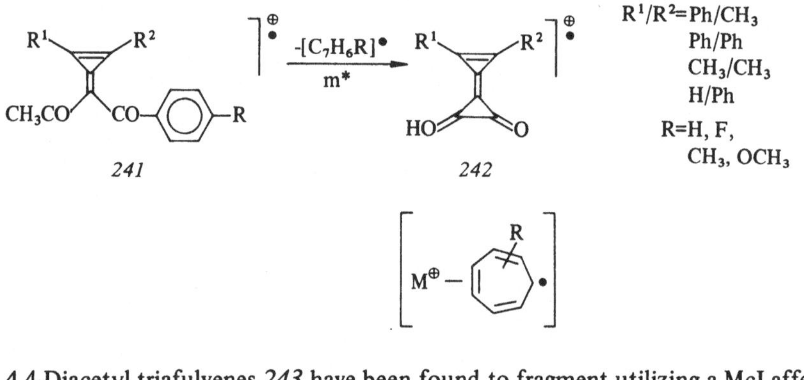

241

242

$R^1/R^2 = Ph/CH_3$
Ph/Ph
CH_3/CH_3
H/Ph

R=H, F,
CH_3, OCH_3

$$\left[M^\oplus - \text{\raisebox{0pt}{\includegraphics{}}}^{\cdot} \right]$$

4,4-Diacetyl triafulvenes *243* have been found to fragment utilizing a McLafferty rearrangement of intermediate *244:*

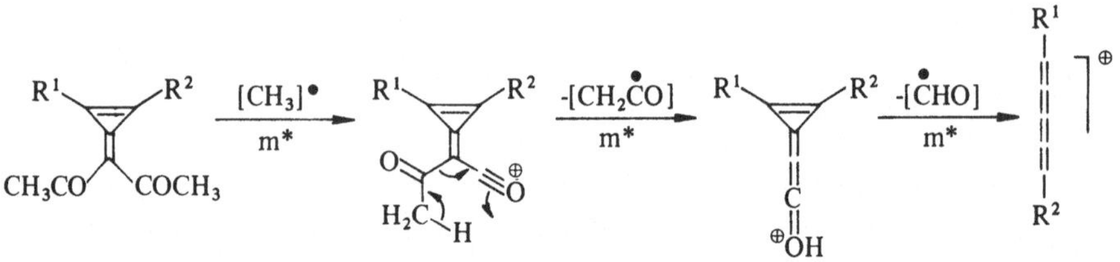

243

244

Interestingly, the fragmentation pattern of the triafulvenes *193/245* dimethyl-substituted at the three-ring ends up into a fragmentation characteristic of a phenyl group, which might well arise from a "randomization" process of C_6 species remaining after loss of the exocyclic substituents according to the following scheme:

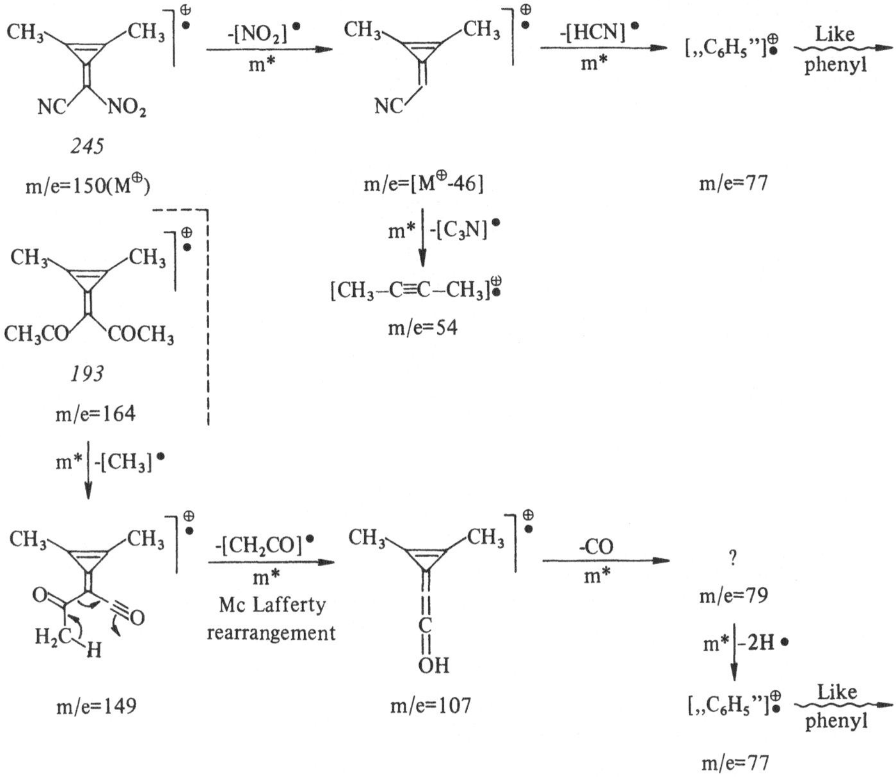

CH₃ ⟍△⟋ CH₃]⊕• $\xrightarrow[m^*]{-[NO_2]^\bullet}$ CH₃ ⟍△⟋ CH₃]⊕• $\xrightarrow[m^*]{-[HCN]^\bullet}$ [,,C₆H₅"]⊕• $\xrightarrow[\text{phenyl}]{\text{Like}}$

NC⟋ ⟍NO₂   NC⟋

245

m/e=150(M⊕)   m/e=[M⊕-46]   m/e=77

$m^* \big\downarrow -[C_3N]^\bullet$

CH₃ ⟍△⟋ CH₃]⊕•   [CH₃–C≡C–CH₃]⊕•

CH₃CO⟋ ⟍COCH₃   m/e=54

193

m/e=164

$m^* \big\downarrow -[CH_3]^\bullet$

CH₃ ⟍△⟋ CH₃]⊕• $\xrightarrow[\substack{m^* \\ \text{Mc Lafferty} \\ \text{rearrangement}}]{-[CH_2CO]^\bullet}$ CH₃ ⟍△⟋ CH₃]⊕• $\xrightarrow[m^*]{-CO}$?

O⟍ ⟍C≋O   C‖   m/e=79

H₂C⟍H   C‖ $m^*\big\downarrow -2H\bullet$

        OH

m/e=149   m/e=107   [,,C₆H₅"]⊕• $\xrightarrow[\text{phenyl}]{\text{Like}}$

                    m/e=77

IV. Reactions of Cyclopropenones and Their Heteroanalogs

1. Thermolysis and Photolysis of Cyclopropenones

a) Decarbonylation

The elimination of carbon monoxide, *i.e.* the (formal) reversal of cyclopropenone formation from divalent carbon species and alkynes, takes place when cyclopropenones are heated to higher temperatures (130–250 °C) or when subjected to photolysis or electron impact[191]:

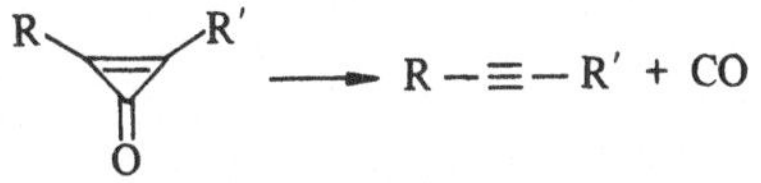

$$R \overset{\triangle}{\underset{O}{\diagup\diagdown}} R' \longrightarrow R-\equiv-R' + CO$$

Generally diaryl cyclopropenones are thermolyzed at lower temperatures than dialkyl cyclopropenones, *e.g.* diphenyl cyclopropenone ~ 150 °C, di-n-propyl cyclopropenone ~ 190 °C, due to the "better" stabilizing properties of alkyl substituents.

Application of the decarbonylation reaction to cyclohepteno cyclopropenone (*39*)[43] led to the intermediate formation of the highly strained cycloheptyne (*246*) as indicated by the formation of its cyclotrimerization product *247* (in analogy to the behavior of dehydrobenzyne[192]) on pyrolysis of *39*; in the presence of a diene reagent, *e.g.* tetracyclone, the product *249* of Diels-Alder addition and subsequent CO elimination was isolated in addition to the spirolactone *250*.

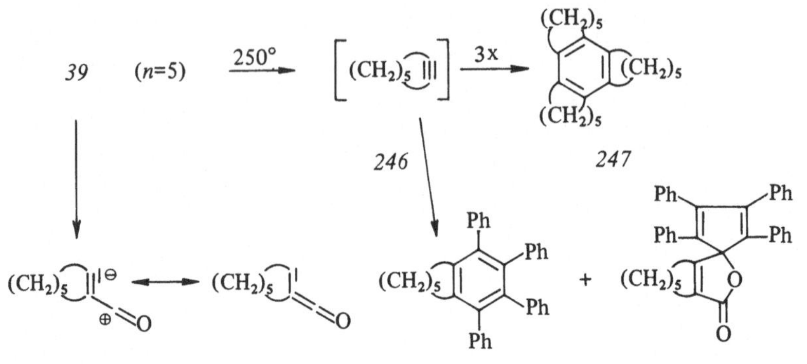

246 *247*

248 *249* *250*

The result of the trapping experiments is also interesting from another point of view. Tetracyclone apparently does not only react with cycloheptyne resulting from cleavage of both C–CO bonds in cyclopropenone *39*, but also with the cyclopropenone itself via the bifunctional valence tautomer *248* resulting from cleavage of only one C–CO bond, since the spirolactone *250* can be derived from a (3 + 2) cycloaddition of *248* to the C=O group of tetracyclone.

The 4-hydroxyaryl substituted cyclopropenone *251* was found by West[193]) to exhibit a remarkable cycle of decarbonylation and oxidation-reduction reactions:

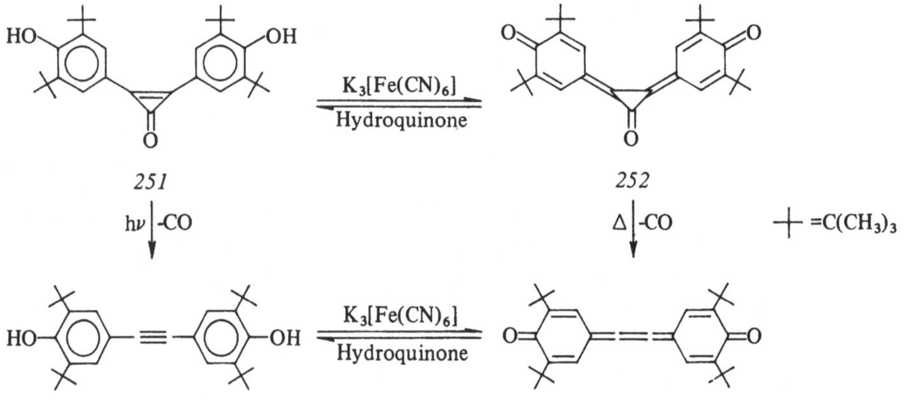

251 *252*

253 *254*

The compound *251* decarbonylates on photolysis to bis(4-hydroxyaryl) acetylene *253*, which is easily oxidized to the quinonoid cumulene *254*. This is also obtained by thermal decarbonylation of the product of oxidation of cyclopropenone *251*, the diquinocyclopropanone *252*. Likewise, the blue derivative of 3-radialene *256* (a phenylogue of triketo cyclopropane) is formed from tris-(4-hydroxyaryl) cyclopropenium cation *255* by oxidation[34]).

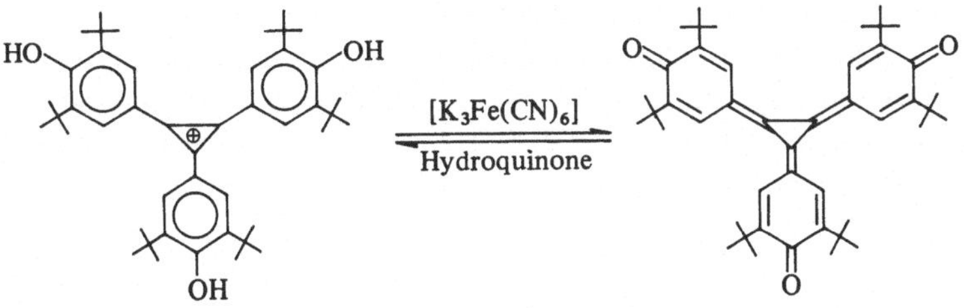

255   256

b) Oligomerization

At temperature below the decarbonylation level cyclopropenones are preferentially transformed to stable dimers, which do not eliminate CO at higher temperatures. Thus, thermolysis of diphenyl cyclopropenone at temperatures above 160 °C gives mainly diphenyl acetylene, whilst heating in the molten state to 145–150 °C[42] or in boiling toluene[194] causes dimerization to spirolactone 257 (R = C_6H_5)[195, 196]. The formation of 257 can again be understood as an addition of one molecule of cyclopropenone through the C^1–CO bond to the C=O group of a second molecule:

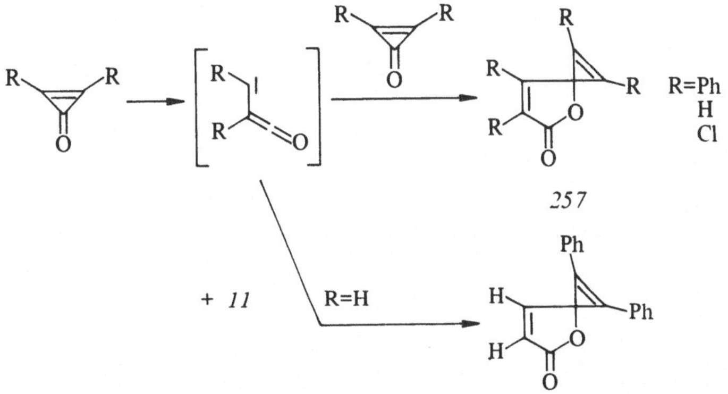

257

258

It should be noted that "codimerization" was achieved from diphenyl cyclopropenone and unsubstituted cyclopropenone (258)[197]. Phenyl hydroxy cyclopropenone, which appears to be an associated dimer in (dioxane) solution[52], formed a dimeric pulviniv acid lactone 260 on treatment with thionyl chloride[51], probably via oxidative rearrangement of a dimer 259:

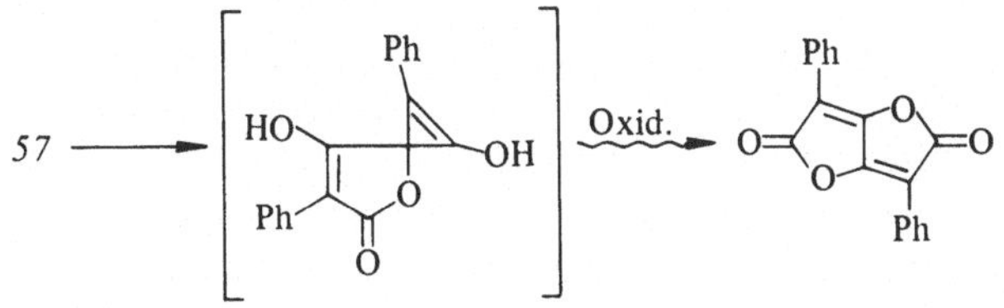

259   260

Another type of dimerization was observed by Japanese authors[198]. In the presence of Ni°, compounds like bis(1,5-cyclooctadiene) nickel(0), diphenyl and di-*n*-propyl cyclopropenone, and cyclohepteno cyclopropenone are transformed to tetrasubstituted *p*-benzoquinones (*261/262*) by formal (2 + 2) or (3 + 3) cycloaddition of two cyclopropenone moieties effected by metal complexing.

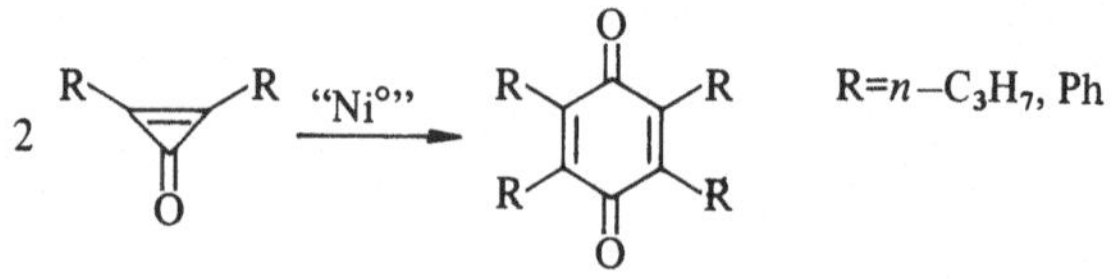

261

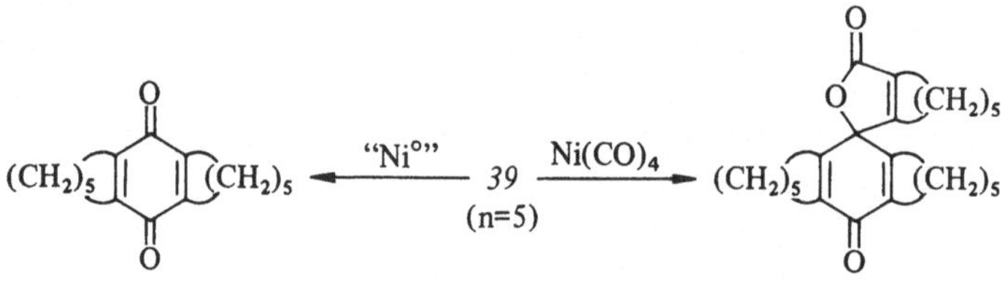

262   *263*

Recently a cyclopropenone trimer was obtained on treatment of cyclohepteno cyclopropenone with Ni(CO)$_4$ in refluxing benzene[199]; structure *263* assigned to this oligomer is derived from addition of a third cyclopropenone molecule to the C=O group of dimer *262*.

Phenyl cyclopropenone is not capable of thermal dimerization. On treatment with Cu^{2+} ions, however, a well-defined tetramer is formed[54], to which structure *265* of a polyene dilactone was assigned. Its generation can be rationalized via *264* with both the above dimerization types contributing in metal-catalyzed reactions.

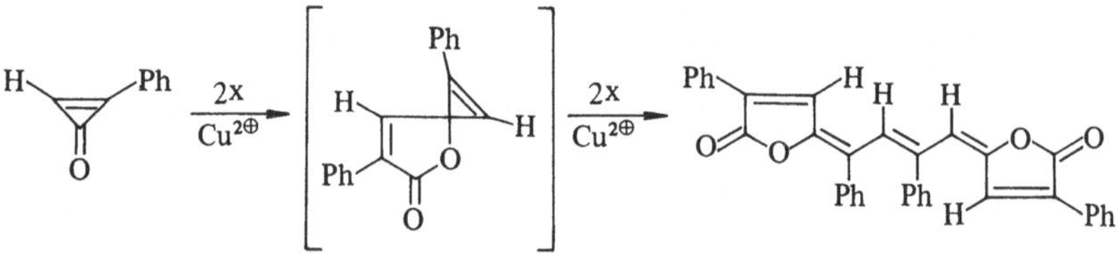

264   *265*

A dimer analogous to *264* has been obtained from methyl cyclopropenone[23] (CH$_3$ instead of C$_6$H$_5$).

Dimerization of cyclopropenones has also been found to occur under reductive conditions. Tetraphenyl resorcinol is formed in addition to a small amount of tetraphenyl *p*-benzoquinone on treatment of diphenyl cyclopropenone with aluminum amalgam[200]; its formation can be rationalized via dimerization of the cyclopropenone ketyl *266* and subsequent aromatization, possibly according to a prismane mechanism.

An analogy at least formally valid can be seen in the reductive conversion of cyclopropenone to hydroquinone occuring with 1 % Na-amalgam[201].

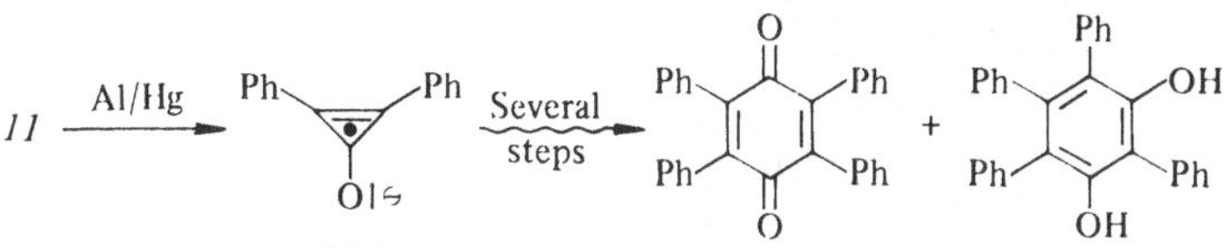

c) Heteroanalogs of Cyclopropenone

Thermolytic and photolytic transformations are reported for several diphenyl cyclopropenone imines and diphenylcyclopropene thione.

Thermal reactions of N-aryl cyclopropenone imines *268* are differentiated by the nature of the N-aryl substituent. Imines *268* (Ar = phenyl, *p*-nitro-phenyl) undergo isomerization to N-aryl-2-phenyl-indenone imines *271* when heated in aprotic solvents[202]. Since in protic solvents, *e.g.* ethanol, only the iminoester *272* is isolated, evidence seems to be given for the intermediacy of *269* implying carbene and ketene imine functionality, which may either cause electrophilic ring closure with a phenyl group to form *271* or may add to the hydroxylic solvent (*272*).

Imines *268* (Ar = phenyl, *p*-anisyl, *p*-chloro-phenyl) are dimerized on heating in aprotic solvents to give the yellow tetraphenyl *p*-benzoquinone imines *267* and (or) their reduction products, the colorless *p*-phenylene diamine derivatives *270*[55].

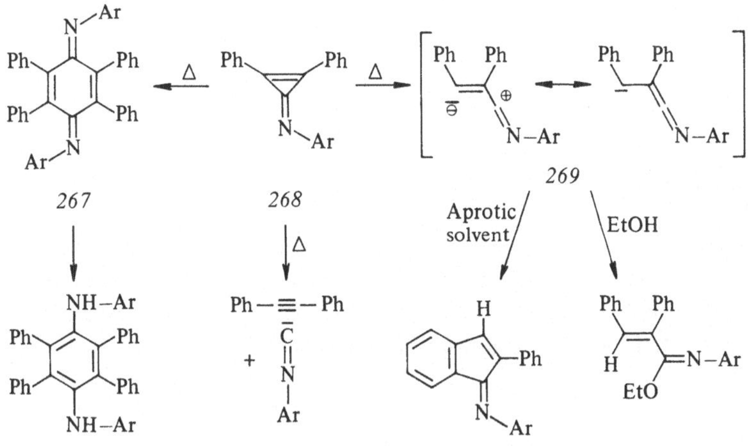

The thermal cycloreversion of imines *268*, *i.e.* formation of isocyanide and alkyne, which would be expected by analogy with cyclopropenone decarbonylation and in reversal of cyclopropenone imine formation (see p. 25), was found to be only a minor side-reaction[203].

Irradiation of diphenyl N(*p*-nitrophenyl) cyclopropenone imine (*268*) gave, in a clear-cut photoreaction, the phenanthreno indenone imine *273*[170] resulting from two units of *268* by loss of two hydrogens and *p*-nitrophenyl isocyanide; the mechanism of this transformation has not yet been elucidated.

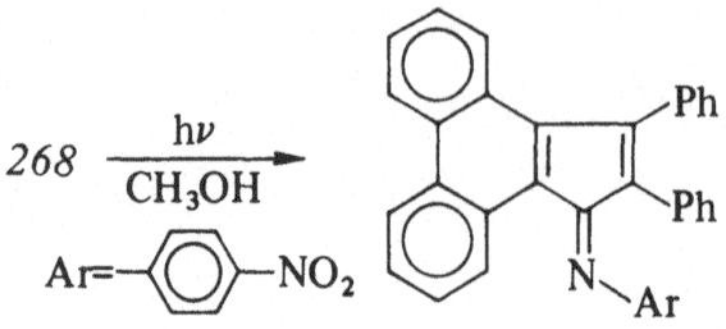

$$268 \xrightarrow[\text{CH}_3\text{OH}]{h\nu}$$

Ar=⟨○⟩—NO$_2$

273

Photolysis of N(p-toluenesulphonyl) diphenyl cyclopropenone imine (*274*) gave rise to a number of products generated by opening of the three-ring as well as by photochemically induced ring closure of the phenyl residues[204]:

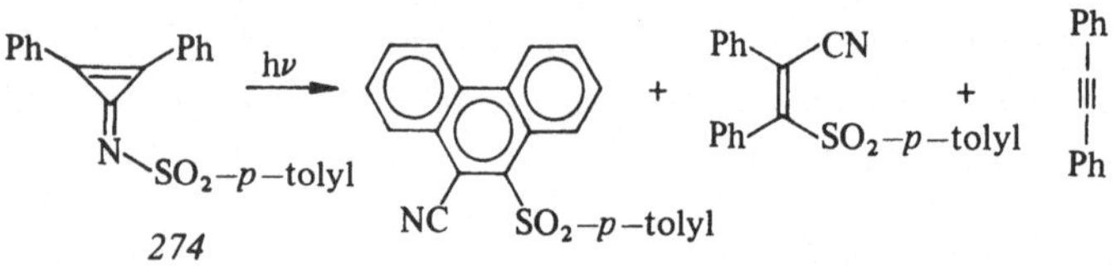

274

Diphenylcyclopropene thione afforded a mixture of dimers on photolysis[205], for which structures *275* and *276* were proposed.

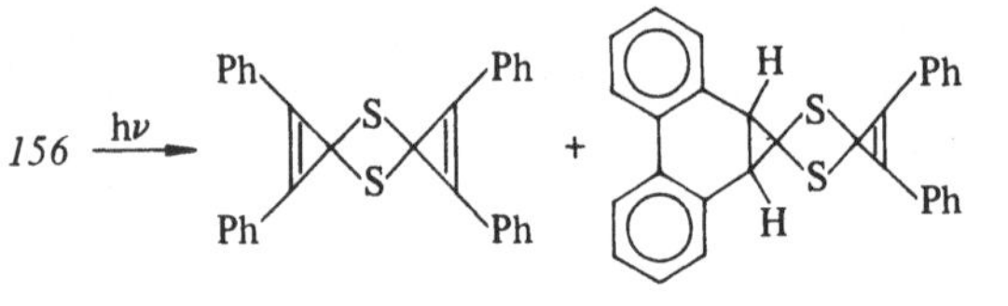

275   *276*

2. Oxidation and Reduction

Cyclopropenones can be oxidized according to several methods. When diphenyl cyclopropenone is treated with alkaline H_2O_2 desoxybenzoin is formed as the main product and is claimed to arise from primary addition of hydroperoxide ion to the C^1/C^2 bond[206]. Treatment with $KMnO_4$ gave benzil[67].

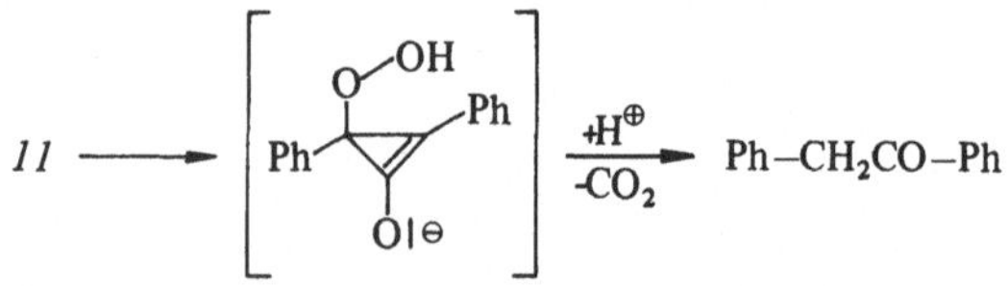

Oxidation of diphenyl or di-tert. butyl cyclopropenone with *m*-chloro peroxybenzoic acid[207] proceeds via intermediates corresponding to a Bayer-Villiger-type mechanism (*277/278*) to "unrearranged" products (1,2-diketones) or "rearranged" products (ketones) depending on the reaction conditions.

Diphenylcyclopropene thione is transformed to diphenyl cyclopropenone by means of lead tetraacetate (presumably via *279/280*), whilst perphthalic acid oxidizes to cation *281*, which gives the unstable S-oxide *282* with $NaHCO_3$[208].

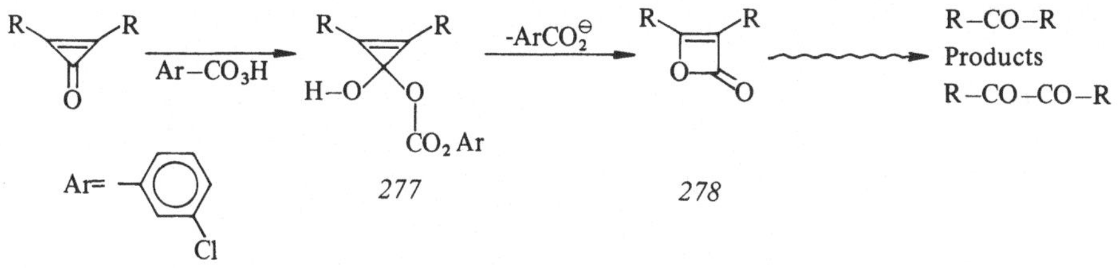

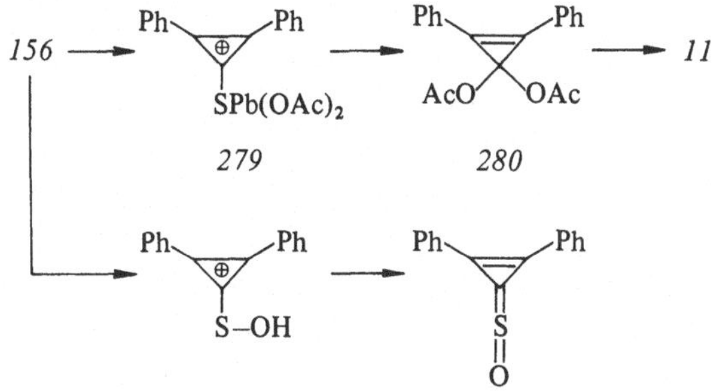

Reduction of cyclopropenones achieved on catalytic hydrogenation varied with the catalyst applied. Thus, with Pt/H_2 from cyclopropenone[28] as well as from di-phenyl[42], di-n-propyl[43] and di-tert. butyl[44] cyclopropenone the acyclic ketones *285* were formed by addition of H_2 to the cyclopropenone C^1/C^2 bond and further reductive cleavage via cyclopropanone intermediate *284*. The same type of three-ring fission was observed on hydrogenation of n-pentyl cyclopropenone[48] with Pd/C catalyst (*286*).

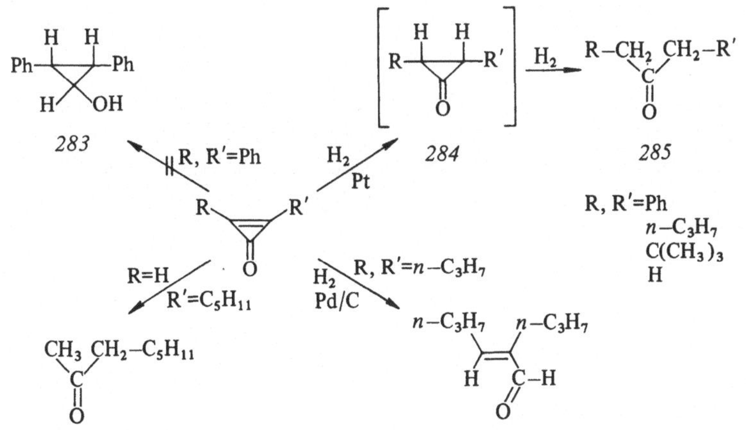

Hydrogenation of di-n-propyl cyclopropenone with Pd/C catalyst, however, gave rise to 2-propyl-2-hexenal (*287*) as a major product according to attack of H_2 at the cyclopropenone C^1/C^3 bond[43]. A cyclopropanone could not be detected spectro-scopically in any case. The formation of diphenylcyclopropanol *283* reported for

the catalytic hydrogenation of diphenyl cyclopropenone[17] has not been substantiated by later investigations[42].

Ethyl phenyl cyclopropenone (*14*) on reduction with NaBH$_4$ gave rise to products *289–291*, which can be ascribed to a common cyclopropanone intermediate
288 ring-opened by further reduction or attack of solvent[209]:

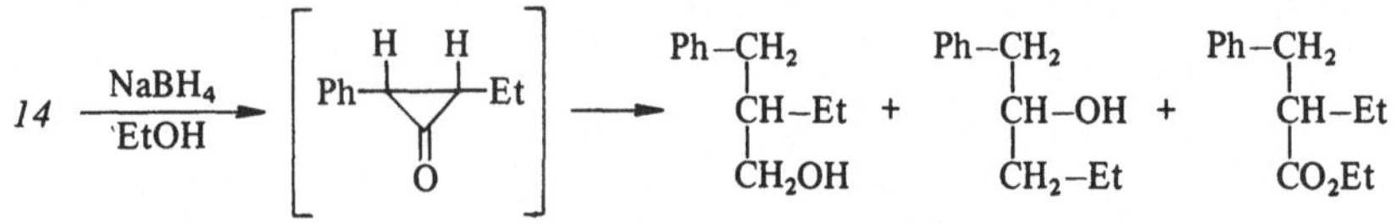

Selective reduction of the cyclopropenone carbonyl group to a CH$_2$ group has
been described for diphenyl cyclopropenone utilizing its protonation product *294*
or the diphenyl chloro cyclopropenium cation *292*, which yielded 1,2-diphenyl-$\Delta^{1,2}$-
cyclopropene (*293*) on treatment with trimethylamine borane[210]:

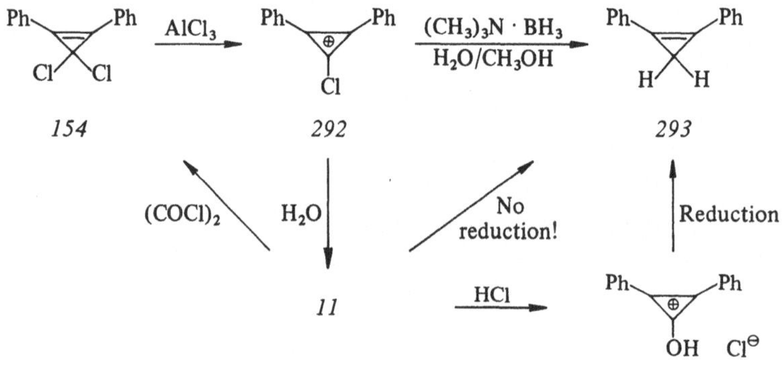

Direct reduction of chloro cation *292* in non-aqueous media to diphenylcyclopropene *293* is possible by means of dimethylamine borane[211]. The chloro cation *292*
is easily prepared from the dichlorocyclopropene *154* and Lewis acids like AlCl$_3$ or
SbCl$_5$[115]. The reductive dimerization of cyclopropenones has already been mentioned (p. 58).

3. Reaction with Electrophiles

a) Protonation, Alkylation, and Acylation

Electrophilic attack on cyclopropenones takes place at carbonyl oxygen, as indicated
by the formation of hydroxy cyclopropenium cations on protonation (see p. 28).
Hydrogen-bonded complexation between the carbonyl oxygen of diphenyl cyclopropenone and the O-H hydrogen in water[212] and substituted acetic acids[213] is reported to give rise to well-defined 1 : 1-adducts (*296*).

Alkylation of cyclopropenones — effected by means of trialkyloxonium tetra-fluoroborates[42, 119] — leads to the easily hydrolyzable O-alkyl cyclopropenium cations 295, which are potential sources for triafulvene synthesis:

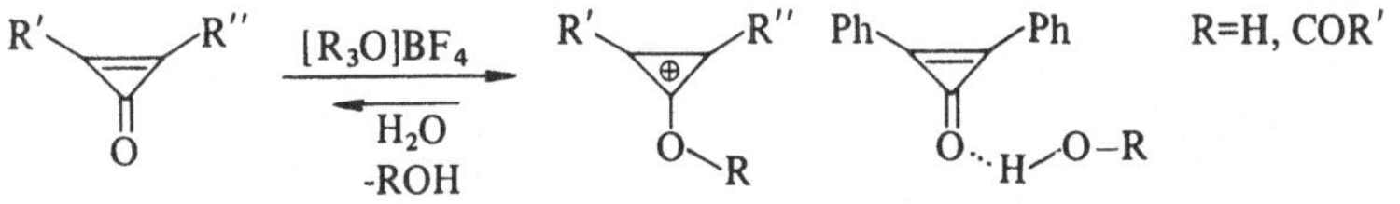

295   296

R=H, COR'

Though O-acyl cyclopropenium cations have not yet been isolated, several cyclopropenone reactions need to be interpreted via intermediary O-acylation.

Diphenyl cyclopropenone is transformed to the dichloride 154 under very mild conditions on treatment with oxalyl chloride[115], a reasonable mechanism implies primary formation of an O-acyl cation 297 suffering fragmentation by loss of carbon monoxide und dioxide:

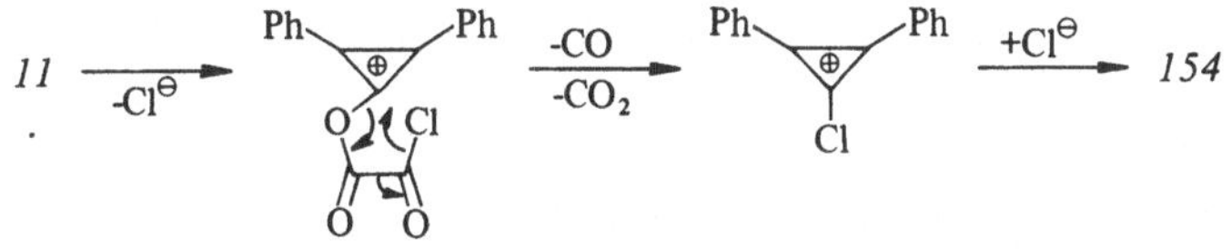

297

Cyclopropenone itself reacts with trifluoroacetic anhydride[197] to the acylal 299, whose formation is easily understood by initial acylation to 298, which subsequently adds trifluoroacetate.

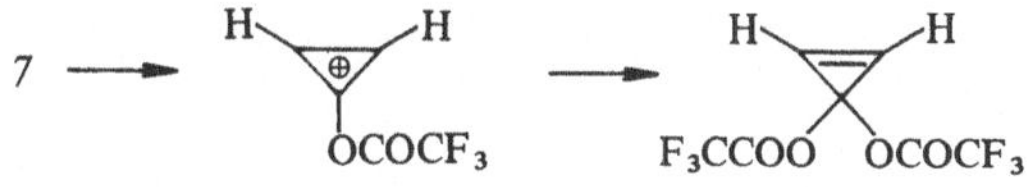

298   299

Likewise, the addition of isocyanates[116] and of diphenyl[117] or bis(trifluoromethyl) ketene[66] to diaryl cyclopropenones is likely to start with an electrophilic attack of the heterocumulene at the carbonyl oxygen. Cycloreversion of the primary adducts 300 and 301 (R = CF$_3$) leads to formation of cyclopropenone imines 152 and triafulvene 63 by elimination of CO$_2$. Analogous formation of tetraphenyl triafulvene (302) from ketene adduct 301 (R = C$_6$H$_5$) does not occur; instead, the naphthol ester 304 is obtained, presumably resulting from 303 by elimination of CO and addition of a second ketene molecule according to the reaction sequence given.

b) Other Electrophilic Reactions

Diphenyl cyclopropenone and dehydrobenzyne react to produce a 1 : 2-adduct, to which the structure 305 of a spirocyclopropene was assigned[214]. Again, its forma-

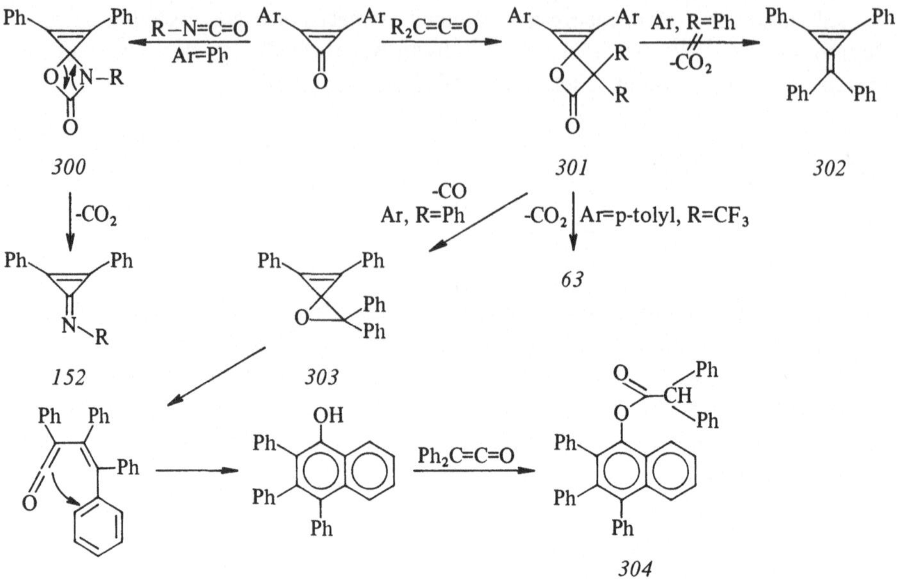

300   301   302

152   303   63

304

tion can be described by a mechanism starting with electrophilic attack of benzyne at cyclopropenone carbonyl oxygen and subsequent valence isomerizations (306–308) finally giving an o-quinocyclopropene, which incorporates the second benzyne moiety by a Diels-Alder addition:

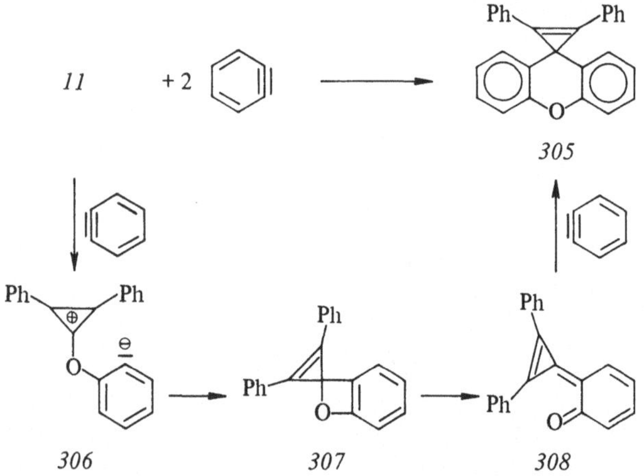

305

306   307   308

Whilst Lewis acids like $SbCl_5$ or $AlCl_3$ form stable adducts with diphenyl cyclopropenone, from which the ketone can be regenerated unchanged[208], trialkyl boranes effect a remarkable ring expansion to 2-phenyl indenone derivatives 309 containing an additional residue in the 3-position[215].

Since the 3-substituent may originate from borane as well as from the solvent [e.g. with triisopropyl borane in diethyl ether indenones 309 were obtained with

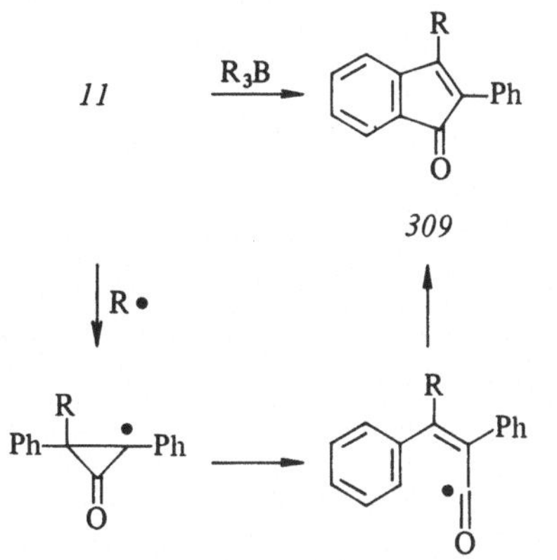

$R = CH(CH_3)_2$ and $-\underset{\underset{CH_3}{|}}{CH}-O-C_2H_5$] a radical mechanism seems to be reasonable to explain indenone formation.

Diborane reacts with diphenyl cyclopropenone[216] to give the product of ring-cleavage *310* after subsequent oxidation. When the oxidative work-up is replaced by protonation, cis-1,2-diphenyl cyclopropane (*311*) is obtained. This indicates that the ring-opening occurs at the oxidation step following the hydroboration of cyclopropene *293*; the primary attack of diborane at carbonyl oxygen (*312*) is supported by experiments with deuterated borane.

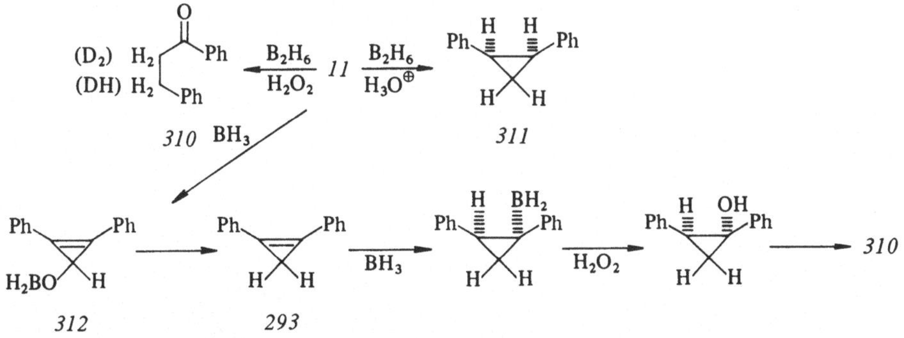

Diphenyl cyclopropenone has been subjected to nitration[217] and bromination[218] in sulphuric acid. According to the second order functionality of the cyclopropenone "substituent" electrophilic substitution of the phenyl residues took place at the *m*-position giving rise to *313*.

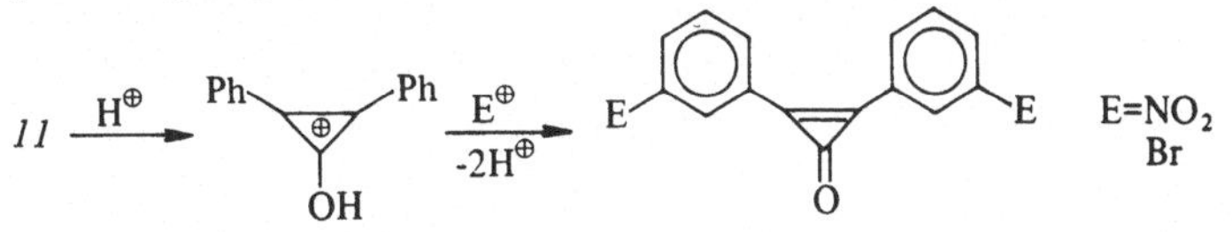

It is interesting that cyclopropenone itself reacts with bromine at low temperatures giving a 1 : 1-compound of probable structure *314*. On warming to 0 °C, *314* is converted to trans-β-bromo acryloyl bromide (*315*)[197]:

$$7 \quad \xrightarrow[-30°]{Br_2} \quad \underset{\overset{OBr}{} \quad Br^{\ominus}}{\overset{H}{\underset{}{\bigvee}}} \xrightarrow{0°} \left[\underset{Br}{\overset{H}{\underset{}{\bigvee}}}_{O-Br} \right] \longrightarrow \underset{O}{\overset{H \quad Br}{Br}}$$

314   315

A mechanistic interpretation is based on the ring-opening principle deduced in the next chapter: the very unusual electrophilic attack of bromine at carbonyl oxygen is followed by nucleophilic addition of bromide ion at elevated temperature and ring-opening by transfer of bromine to C^1/C^2.

4. Reactions with Nucleophiles

a) Cyclopropenones

Cyclopropenones interact with hydroxide or alkoxide ions[42, 43], primary and secondary amines[209, 219], hydrazine[124], carboxylates and amides[220] as well as thioamides[221] forming acrylic acids and their derivatives. In a general scheme, primary addition of the nucleophile at carbonyl carbon (C^3) is followed by stabilization of the intermediate 316 by ring opening to anion 317 and proton transfer to $C^1(C^2)$:

$$\underset{O}{\overset{R}{\underset{1}{\bigvee}}}\overset{2}{\underset{3}{R}} \xrightarrow{Nu-H} \underset{\overset{I}{O}_{\ominus} \quad \overset{Nu-H}{\oplus}}{\overset{R}{\underset{}{\bigvee}}}R$$

316

$$\underset{\overset{Nu-H}{\oplus}}{\overset{R \quad R}{O=}} \longrightarrow \underset{Nu}{\overset{R \quad R}{O=}}_{H}$$

317

Nu = OH, OR, OCOR
NHR, NR$_2$, NHNHR
NH–CS–R

The reaction with hydroxide ion is frequently used as proof for the chemical structure of cyclopropenones and has been examined in some detail with respect to the factors governing ring-cleavage. Thus, methyl cyclopropenone[23] and aqueous NaOH react to yield a mixture of methacrylic and crotonic acids in a ratio of 3 : 1 as expected from the relative stabilities of the two possible intermediate carbanions (type 317):

$$\underset{O}{\overset{R}{\underset{}{\bigvee}}}R' \xrightarrow{OH^{\ominus}} \underset{H \quad COOH}{\overset{R \quad R'}{}} + \underset{HOOC \quad H}{\overset{R \quad R'}{}}$$

318   319

Phenyl aryl cyclopropenones[16] were cleaved by methanolic KOH to a mixture of cis aryl cinnamic acids ($318/319$; R = phenyl, R' = aryl), whose rates of formation gave rise to a linear Hammett-type correlation with σ values in the range of -0.268 to $+0.373$ and $\rho = 0.75$. This also indicates that cleavage yielding the more stable carbanion is preferred. Interestingly, ortho-substituted aryl phenyl cyclopropenones gave only α-phenyl-β-aryl acrylic acids (319; R' = phenyl, R' = aryl), which was ex-plained in terms of steric interactions.

In contrast to the above general scheme (p. 66), the strained bicyclic cyclopro-penone 40[47] is attacked by hydroxide ion at *both* three-ring sites (C^3 and $C^{1(2)}$) resulting in formation of acid 320 and 1,2-diketone 321 in a ratio of 2 : 3,

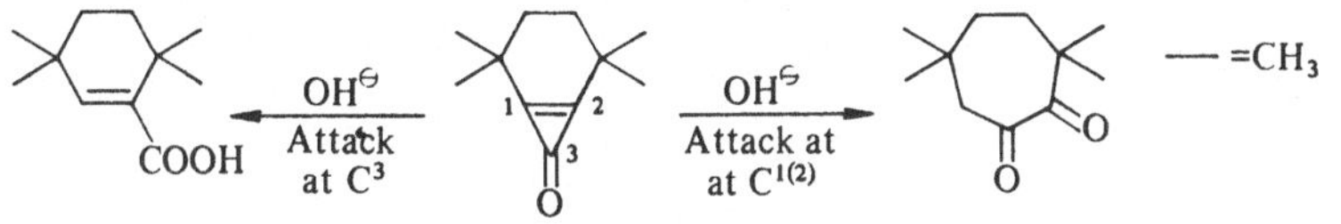

whilst from cyclohepteno cyclopropenone only the product of C^3-attack (cyclo-heptene-1-carboxylic acid) is produced with alkali[43] (see also p. 48) and Ref.[209]

A rather complex reactivity towards the cyclopropenone system is exhibited by N-nucleophiles. Thus, ammonia reacts with diphenyl cyclopropenone to yield either the enamino aldehyde 323[222] or a mixture of the cis and trans isomeric diphenyl azetidinones 322[223] depending on the reaction conditions; these products result from primary addition of the nucleophile at $C^{1(2)}$:

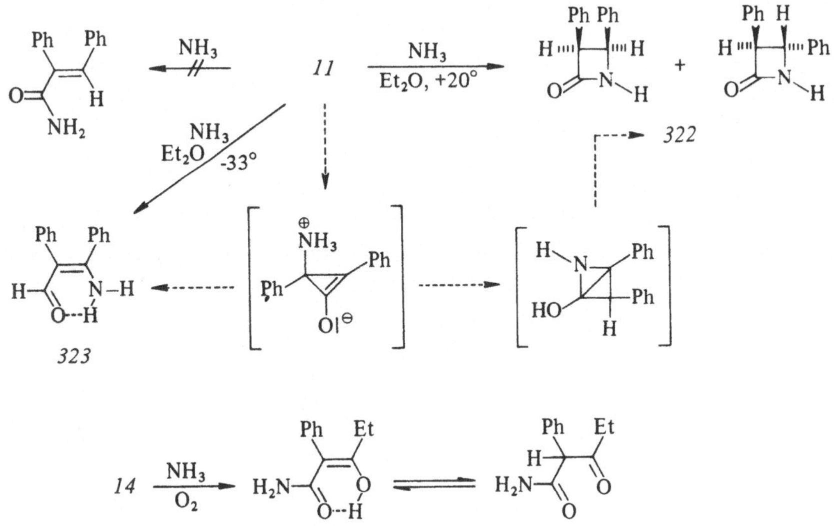

However, ethyl phenyl cyclopropenone (14) was found to give the β-keto amide 324 with ammonia in the presence of oxygen[209] and this demands primary attack of NH_3 at C^3. Apparently product formation is influenced — in a fashion not yet com-

67

pletely understood — not only by the nature of substrates (substitution and steric requirements), but also by the nature of the solvent employed (proticity and polarity).

This is further accentuated by the surprising results of the reaction between aziridine and diphenyl cyclopropenone which was elucidated by Dehmlow[224]. In aprotic media two molecules of aziridine react with a cyclopropenone moiety eliminating ethylene and forming enamino amide *327*, whereas in protic media one molecule of aziridine reacts with the exclusive formation of the aziridide *326*:

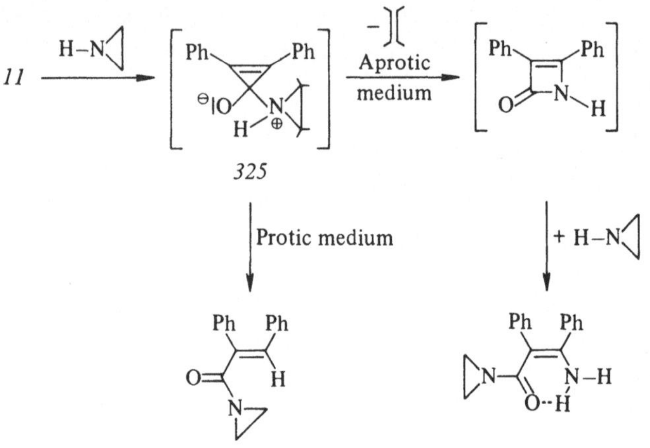

These findings can be interpreted in terms of a "normal" ring-opening mechanism of intermediate *325* with proton transfer favored by protic solvent, whilst in aprotic solvent cycloreversion of the unstable aziridinium grouping in *325* followed by ring expansion prevails. Likewise, 2,3-disubstituted aziridines follow this reaction pattern, while N-substituted aziridines do not[225].

Amongst other N-nucleophilic species, hydroxylamine exhibits some abnormal behavior besides oxime formation (p. 25). Thus it reacts with diphenyl cyclopropenone[42] probably by 1,4-addition and subsequent oxidation and/or decarboxylation giving rise to 3,4-diphenyl isoxazolone (*328*) and desoxybenzoin oxime. With pentyl cyclopropenone[48] hydroxylamine undergoes addition followed by "normal" oximation after ring fission yielding 2,3-dioximino octane (*329*).

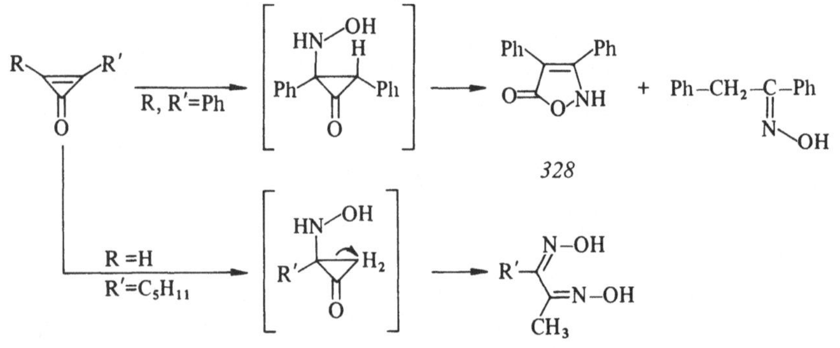

Recently, N-aryl sulphimides were found to react with diphenyl cyclopropenone and its thione[226] giving rise to imines *331* and *332*. Apparently attack of the sulphimide at cyclopropenone C^3 gives rise to intermediacy of a ketene *330*, which is consumed by an excess of the sulphimide (to give *331* after hydrolysis) or by its Sommelet-Hauser rearrangement (to give *332*):

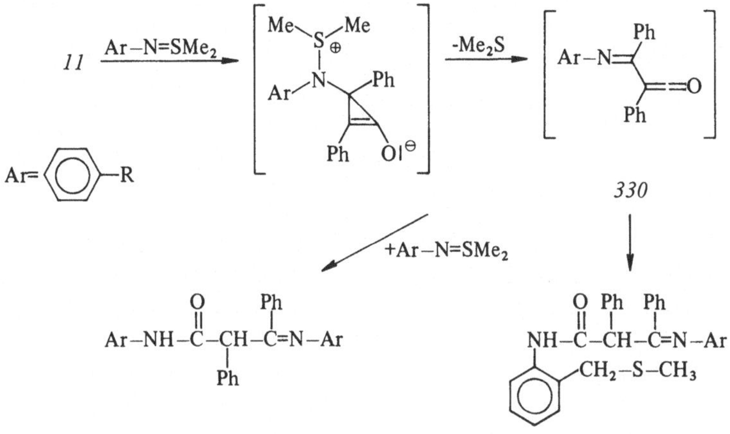

331   *332*

Intramolecular interception of the ketene intermediate by an internal nucleophile (as available *e.g.* in the sulphimide *333* derived from 2-aminopyridine) gave rise to annelated pyrimidones, *e.g. 334:*

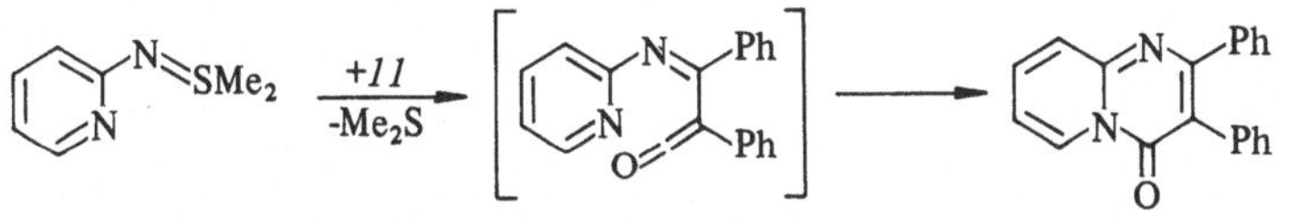

333   *334*

Organometallic compounds react with cyclopropenones either at C^3 or at $C^{1(2)}$ position depending on metal component and ring substitution. Thus the addition of Grignard's reagents to diphenyl cyclopropenone[42, 57] or of organolithium compounds to di-tert. butyl cyclopropenone[44] yields cyclopropenium cations *335/336:*

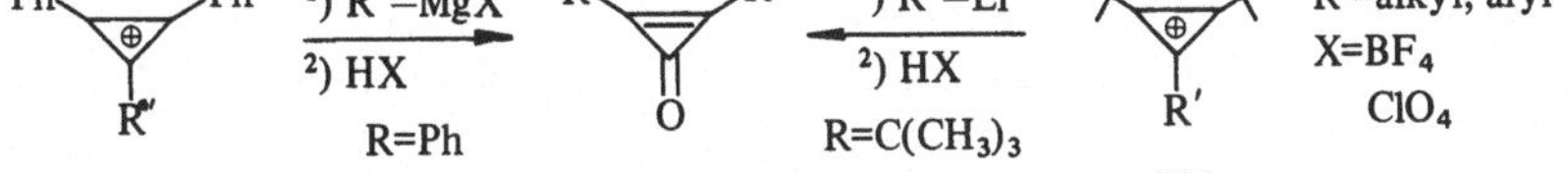

335   *336*

However, diphenyl cyclopropenone undergoes conjugate addition via $C^{1(2)}$ with organolithium reagents[227, 228] as indicated by products of structure *337* and *338:*

69

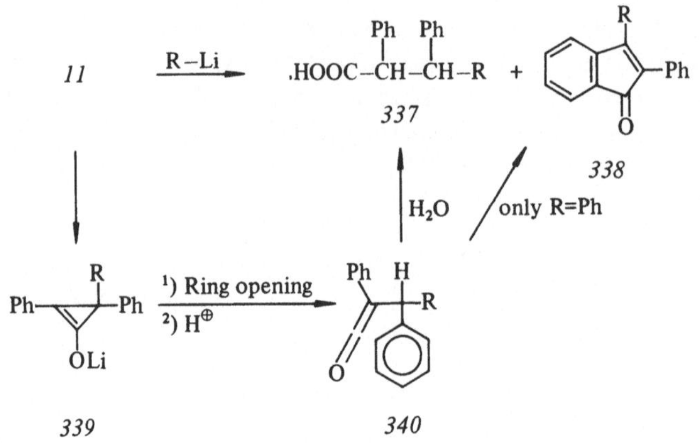

A reasonable mechanism for their formation starts with the primary adduct *339*, which is capable of ring-opening to the ketene *340;* this can either be trapped by addition of water (*337*) or undergo intramolecular acylation followed by dehydrogenation (*338*).

Similiarly, cyclopropenone reacts with Grignard's compounds via conjugate addition, which is followed by an "ene"-reaction of intermediate *341* with a second cyclopropenone moiety (*342*) leading to 2-substituted resorcinols[201].

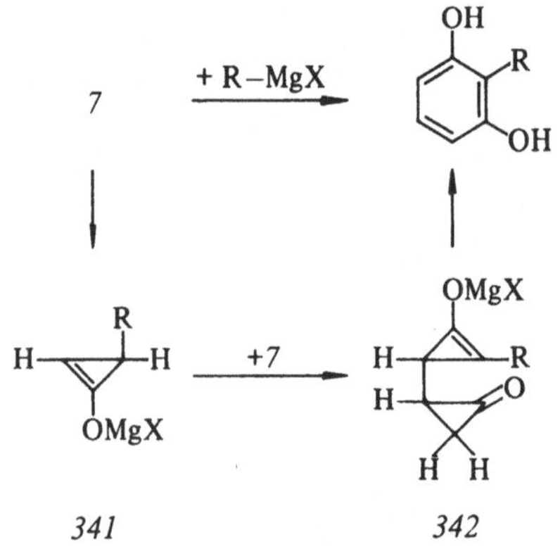

Finally, a reaction should be mentioned in which a nucleophile gives "support" to another reacting species without appearing in the final product. Diphenyl cyclopropenone interacts with 2,6-dimethyl phenyl isocyanide only in the presence of triphenylphosphine with expansion of the three-ring to the imine *344* of cyclobutenedione-1,2 [229, 230]. Addition of the isocyanide is preceded by formation of the ketene phosphorane *343*, which can be isolated in pure form[55, 231]; it is decomposed by methanol to triphenyl phosphine and the ester *52*.

Phosphorane *343* is interesting from another point of view as it represents a formal "trapping" product of the species *345* resulting from cleavage of one C–CO bond in cyclopropenone claimed earlier (p. 56).

70

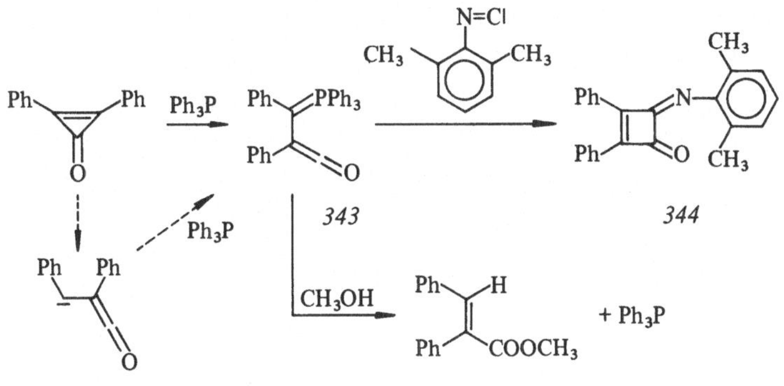

343

344

345

52

b) Heteroanalogs

Diphenyl cyclopropenone imines[88], hydrazones[232], oxime[232], and diphenylcyclopropene thione[219] as well as cyclopropenium immonium cations[88] were found to
undergo facile ring-opening reactions with amines. The imine *268* and the oxime
346 are attacked by secondary amines at $C^{1(2)}$ giving rise to the vinylogous amidines
347 and the enamino nitriles *348*, respectively:

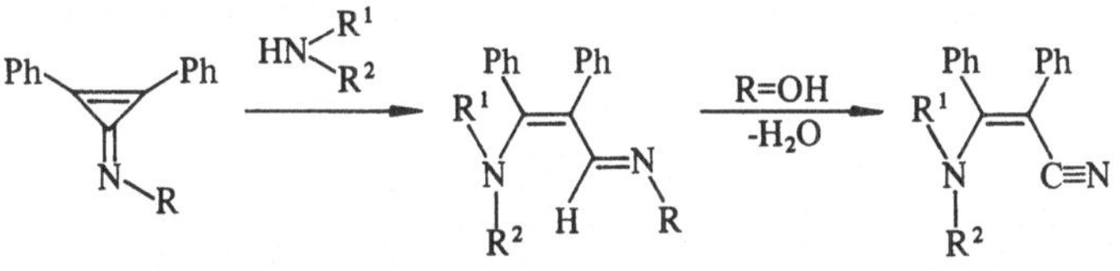

268 : R=Ar
346 : R=OH

347

348

Thione *156* and the hydrazones *349* are cleaved preferentially via attack of the
nucleophile at C^3 of the cyclopropenone system yielding products *350*:

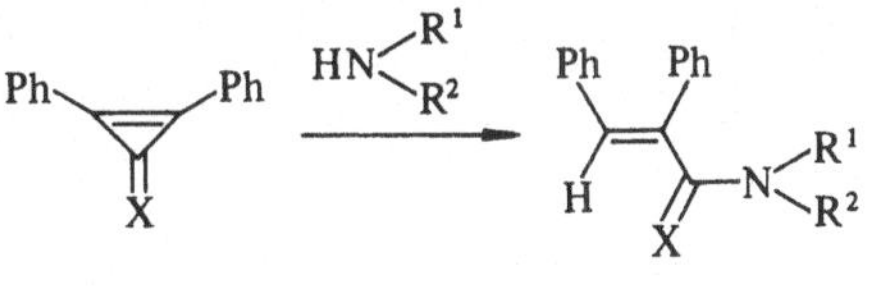

156 : X=S
349 : X=N–NHR

350

The products from the cyclopropenium immonium cations *153* and primary and
secondary amines[88] vary with amine structure and basicity of the amino function

attached to the three-ring. With weakly basic immonium groupings, *e.g.* $\overset{\oplus}{=}N\overset{\diagup CH_3}{\diagdown Ph}$,

exchange of amine residue dominates, whereas increase of immonium group basicity

71

gives rise to nucleophilic ring cleavage starting at $C^{1(2)}$ and C^3 and affording amidinium cations *351/352:*

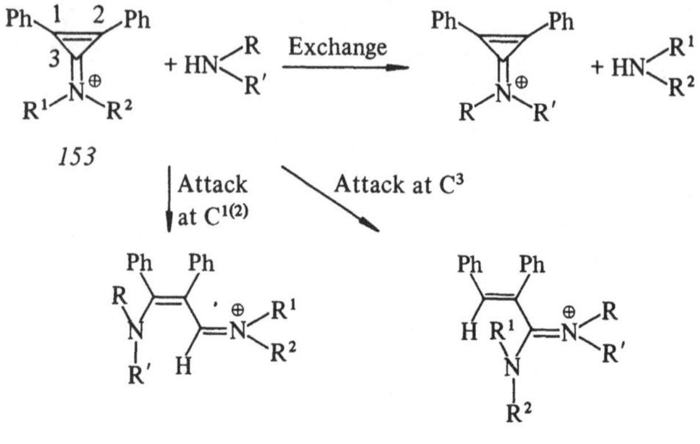

153

Attack at $C^{1(2)}$ — Attack at C^3

352

5. Reactions with Systems Containing Multiple Bonds

a) Diels-Alder Reactions

In general, cyclopropenones do not exhibit dienophilic qualities despite their highly strained C^1–C^2 "double" bond. An exception is made by unsubstituted cyclopro-

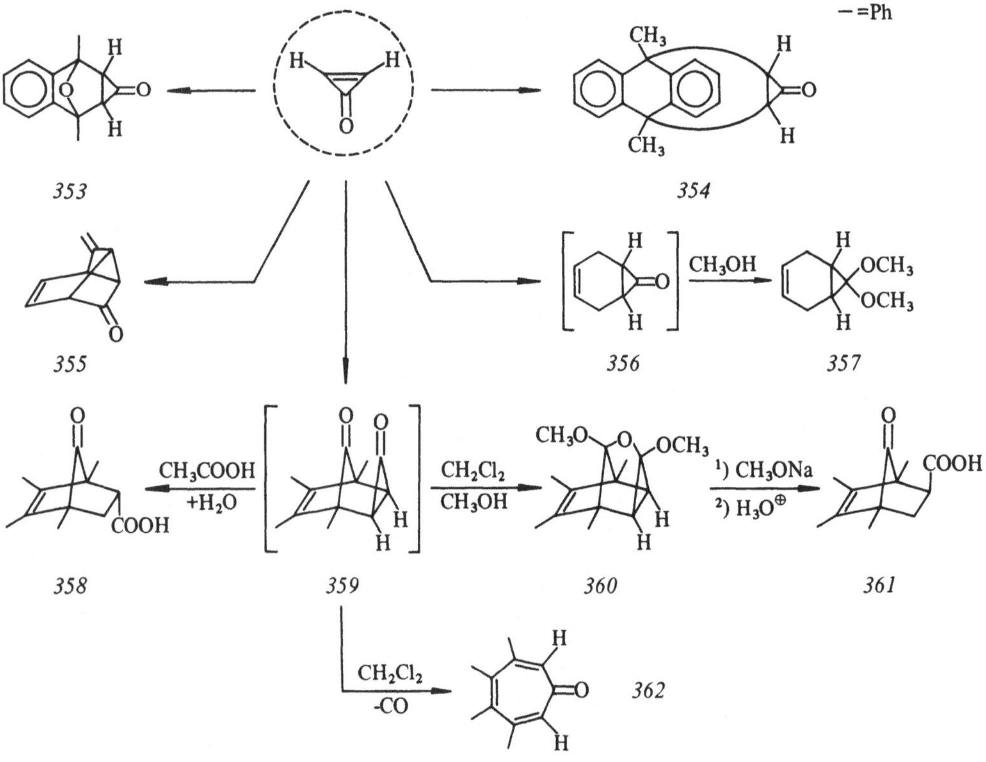

$- = Ph$

353

354

355

356

CH₃OH

357

358

CH₃COOH
+H₂O

359

CH₂Cl₂
CH₃OH

360

1) CH₃ONa
2) H₃O⊕

361

CH₂Cl₂
-CO

362

penone, which was found[201] to undergo a series of "classical" (2+4) cycloaddition of the Diels-Alder type.

9,10-Dimethyl anthracene and diphenyl isobenzofuran form remarkably stable[233] cyclopropanone derivatives (353/354), whilst with other diene components (butadiene, tetracyclone, and fulvene) the primarily formed Diels-Alder adducts either suffer ketalizing attack of the solvent (356 → 357, 359 → 358/360) or undergo irreversible changes such as decarbonylation to 362 or rearrangement to 355.

Diphenyl cyclopropenone has been reacted with activated dienes like diphenyl isobenzofuran[234] and diethylamino butadiene[194]. Interestingly, the isobenzofuran does not add to the C^1/C^2 bond as in cyclopropenone, but to the C^1/C^3 bond in a (3+4) cycloaddition as indicated by the product 363.

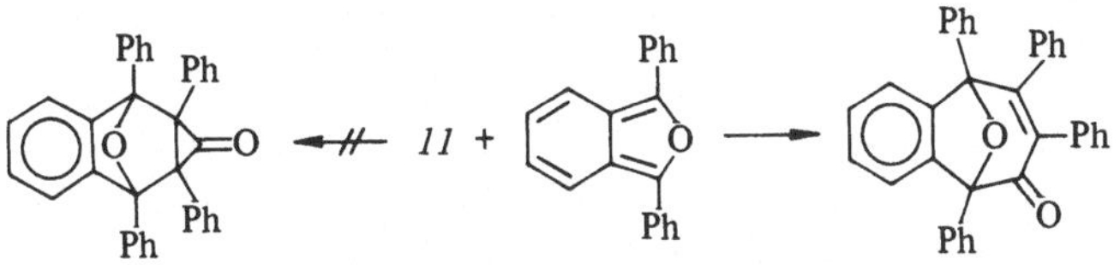

363

The dieneamine, however, maintains the "normal" (2+4) mode of diene reactivity followed by elimination of diethyl amine, thus yielding 2,7-diphenyl tropone (364).

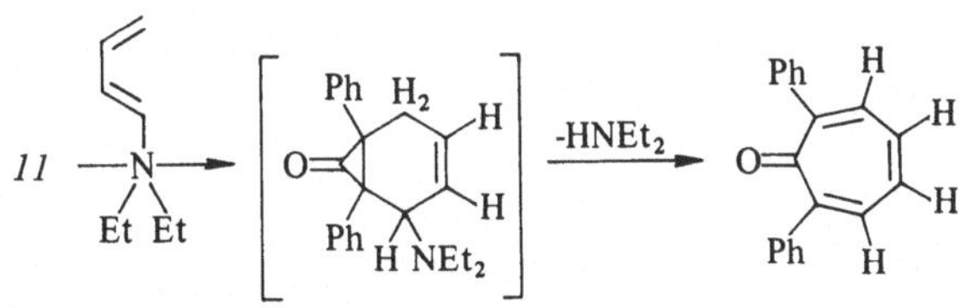

Analogous dienophilic behavior is shown by diphenyl cyclopropenone towards oxazoles[235] (isoxazoles behave quite different, see p. 78) giving rise to 3-ethoxy-2,6-diphenyl pyrone-4 (366), which may result from a primary Diels-Alder adduct 365 stabilized by elimination of nitrile:

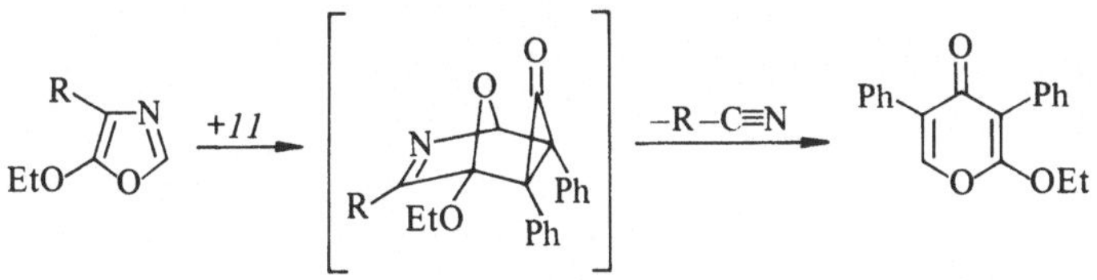

365 366

b) Reactions with Electron-rich Multiple Bonds (Enamines, Ketene Acetals, and Ynamines)

According to earlier reports diphenyl cyclopropenone and enamines[194] or ketene acetals[236] form the cross-conjugated β-aminoenones 368, which were assumed to

73

result from a $(2+2)$ cycloaddition of the enamine double bond to the C^1/C^2 bond of the cyclopropenone system via a dipolar intermediate *367*; this intermediate has been isolated in the case of ketene acetals[236].

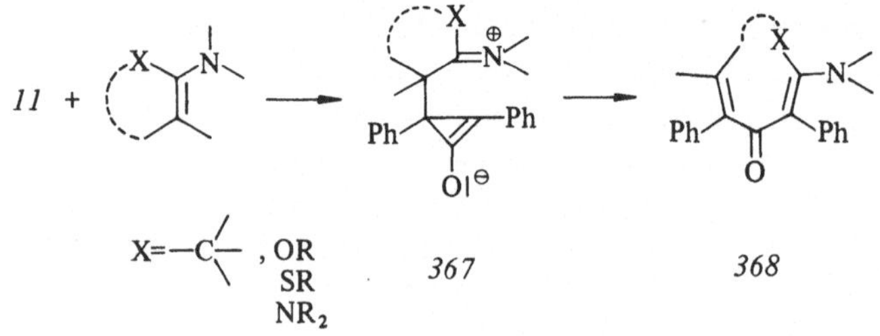

The above structures, however, and the resulting mechanistic consequences were re-evaluated by Dreiding[237-239], Sauer[240] and others[241-244].

At low temperatures cyclopropenones and enamines or ketene acetals were shown to yield 2-azonia-bicyclo(3,1,0)hex-3-enolates-3 (*371*, X=O), which can be isomerized thermally to penta-2,4-diene amides (*372*, X=O). At elevated temperatures the amides were found to be the principal products arising from "C-N-insertion"[237] (insertion of the cyclopropenone three-carbon unit into the C-N bond of the enamine). These were accompanied in some cases by β-aminoenones *373* arising from "C-C-insertion" [237] (insertion of the cyclopropenone into the C-C double bond of the enamine) and α-amino cyclopentenones *375* formed by Stevens rearrangement of the ylide *369* and cyclopentenones *374* ("condensation"[237]).

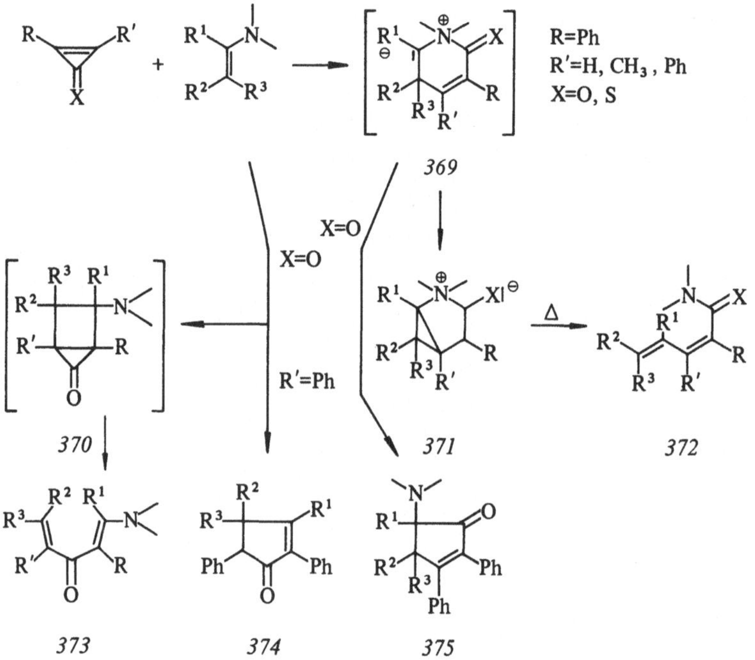

Diphenylcyclopropene thione shows analogous reactivity[245,246] forming 2-azonia-bicyclo(3,1,0)hex-3-enethiolates-3 (*371*, X=S) at low temperatures and thioamides (*372*, X=S) at elevated temperatures.

In the reaction scheme (formulated above for enamines) the primary formation of an "acyl ylide" *369* (the formal product of addition of the enamine sequence C=C-N to the C^1/C^3 bond of cyclopropenone) was first suggested by Dreiding[237]. This was confirmed by findings on diphenylcyclopropene thione[246], which gave a mixture of syn and anti stereoisomeric betaines *377/378* when reacted with enamine *376* possessing exclusively a Z-configuration:

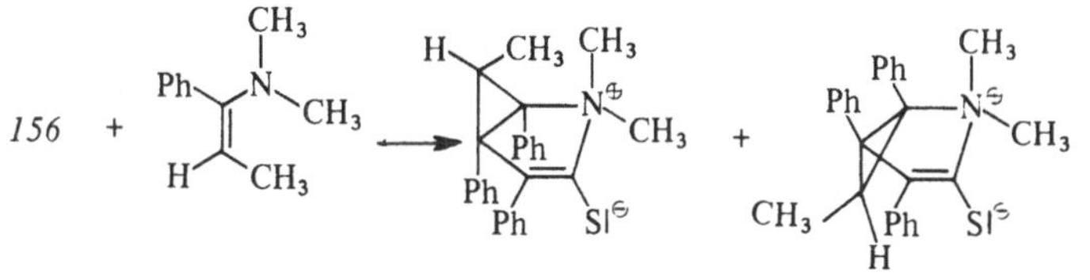

156 + 376 → 377 + 378

Diphenylcyclopropene thione also reacts with Schiff bases *379* via the tautomeric vinylamine form to give betaines *380*, which tautomerize to the bicyclic thioamides *381*[247]:

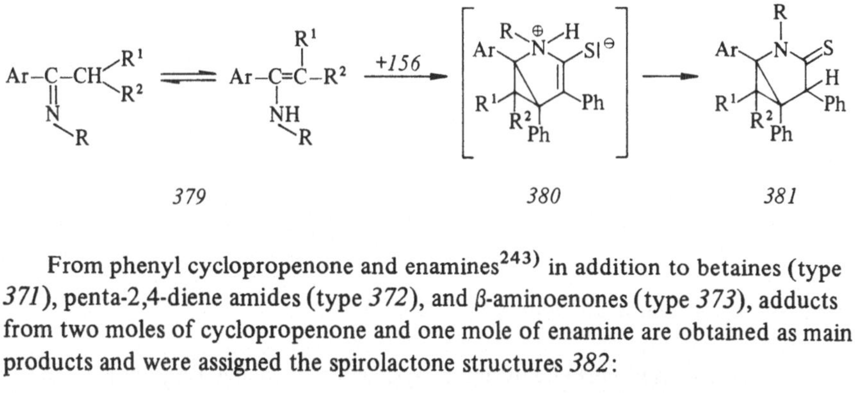

379 +156 → 380 → 381

From phenyl cyclopropenone and enamines[243] in addition to betaines (type *371*), penta-2,4-diene amides (type *372*), and β-aminoenones (type *373*), adducts from two moles of cyclopropenone and one mole of enamine are obtained as main products and were assigned the spirolactone structures *382*:

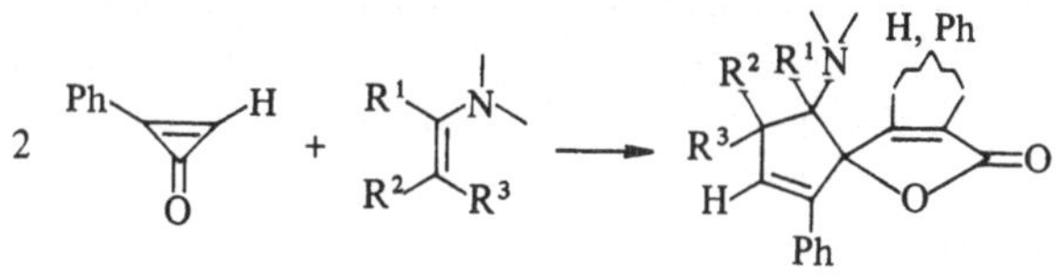

382

The bicyclic enamine *383* deviates from the above reaction scheme: when interacting with diphenyl cyclopropenone the betaine *384* formed initially does not isomerize to the amide *385*, but to the α-amino cyclopentenone *386*, possibly favored by steric reasons[248].

Diphenyl cyclopropenone also has been reacted with ynamines, e.g. *387*[249]. Since cyclopentene dione *389* was obtained after hydrolytic work-up, amino cyclo-

75

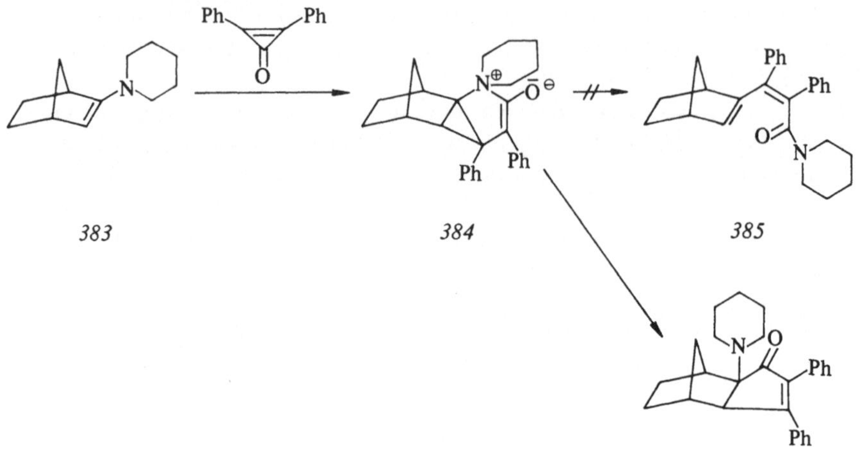

383   384   385

386

pentadienone *388* was thought to be the primary reaction product. It cannot be concluded on the basis of these results whether *388* resulted from an addition of the electron-rich triple bond to the C^1/C^2 or the $C^{1\,(2)}/C^3$ bond of cyclopropenone.

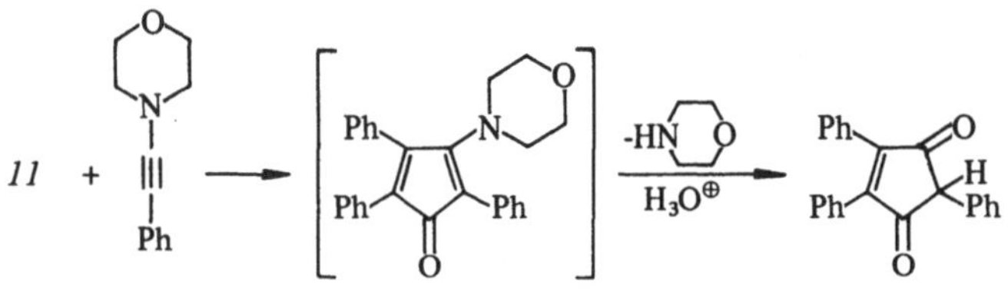

387   388   389

c) Reactions with Systems Containing C=N Bonds

N-Heteroaromatic compounds like pyridine, pyridazine, pyrazine, isoquinoline, and
their derivatives[42,250] react with diphenyl cyclopropenone in a formal (3+2) cycloaddition mode to the C=N bond of the heterocycle. As expected from the results discussed earlier (p. 67), the reaction is initiated by attack of nitrogen at the cyclopropenone C^3 position and followed by stabilization of the intermediate betaine *390*
through nucleophilic interaction of the C^1/C^3 bond with the activated α-site of the
heterocycle, giving rise to derivatives of 2-hydroxy pyrrocoline (*391–394*). In some
cases, e.g. diphenyl cyclopropenone and pyridine[42], further interaction with a second
cyclopropenone molecule is possible under the basic conditions leading to esters of
type *392*.

Cinnoline[250], which incorporates diphenyl cyclopropenone to give N-N annellated
product *395*, is an exception.

Systems such as Schiff bases interact with cyclopropenones in a different way
from the above scheme. Thus, ketimines and diphenyl cyclopropenone afford Δ^2-

76

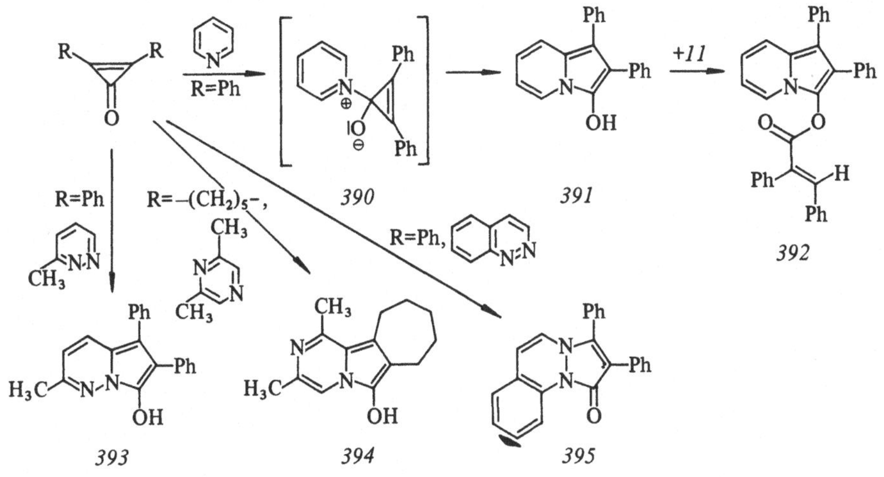

pyrrolin-4-ones $396^{251)}$, which should arise from $(3+2)$ cycloaddition to the $C^{1\,(2)}/C^3$ bond of the cyclopropenone initiated by attack of the azomethine nitrogen at $C^{1\,(2)}$:

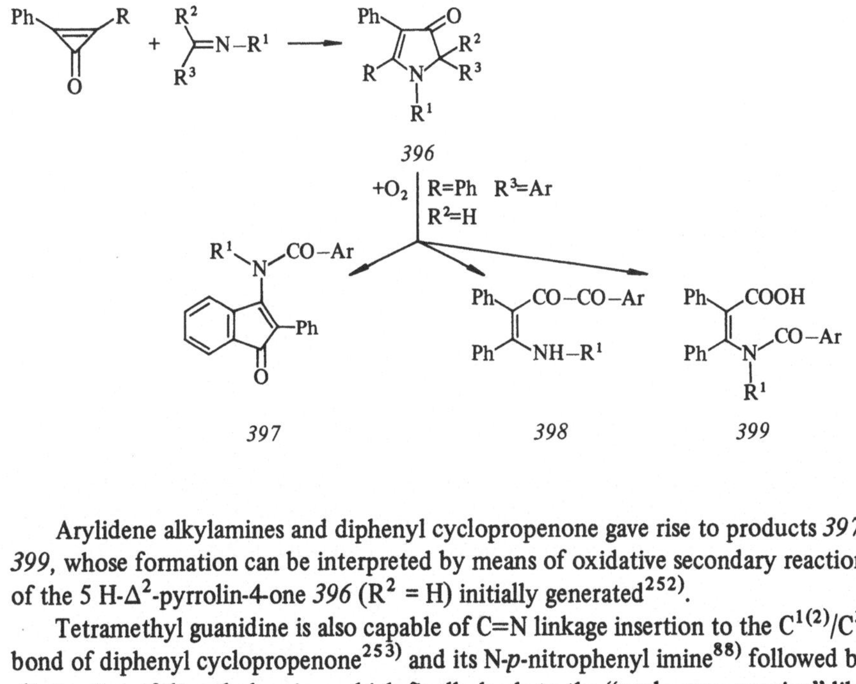

Arylidene alkylamines and diphenyl cyclopropenone gave rise to products *397–399*, whose formation can be interpreted by means of oxidative secondary reactions of the 5 H-Δ^2-pyrrolin-4-one *396* (R^2 = H) initially generated$^{252)}$.

Tetramethyl guanidine is also capable of C=N linkage insertion to the $C^{1(2)}/C^3$ bond of diphenyl cyclopropenone$^{253)}$ and its N-*p*-nitrophenyl imine$^{88)}$ followed by elimination of dimethyl amine, which finally leads to the "cyclo-merocyanine"-like 3-azacyclopentadienone derivatives *400*.

Azirines$^{254)}$ in principle react analogously to Schiff bases: the 2,3-diphenyl pyridones-4 *403* obtained from diphenyl cyclopropenone may well result from a primary betaine *401*, which reorganizes to the pyridone-4-system via its valence tautomer, the ketene *402*.

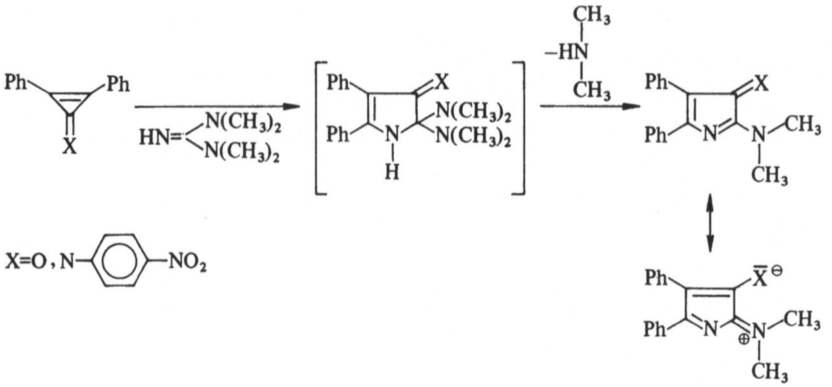

400

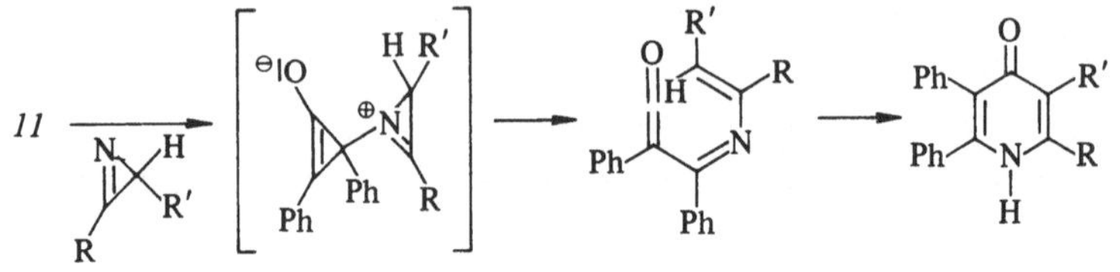

X=O, N—⟨◯⟩—NO₂

401   402   403

4-Pyridones are formed from diphenyl cyclopropenone and isoxazoles[255] in a similiar pathway:

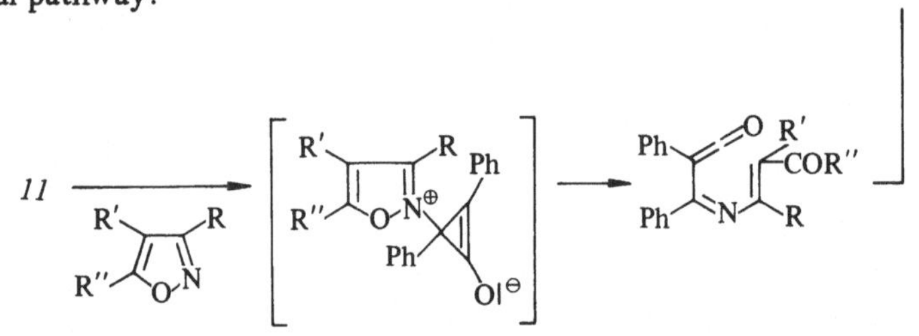

It should be noted that the isopyrazole 404 reacts with diphenyl cyclopropenone in a rather complicated fashion[256]. The products of assigned structures 405/406 may stem from addition to the cyclopropenone C^1/C^2 bond and to the $C^{1(2)}/C^3$ bond.

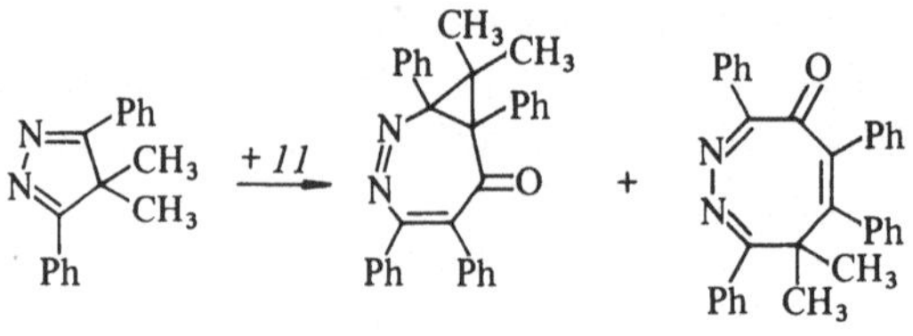

405   406

6. Reactions with 1,3-dipoles

Diazoalkanes like diazomethane, -ethane and -propane react with diphenyl cyclopropenone to give 3,5-diphenyl-4-pyridazinones 408[42, 257]; 2-diazopropane, however, yields the diazoketone 410[258]. Since in these products the phenyl-bearing cyclopropenone carbons are separated, the decisive reaction step is likely to be cycloaddition of the 1,3-dipole to the three-ring C^1/C^2 bond, subsequent ring opening of the intermediate bicyclopropanone 407 being determined by the substituents at the diazo carbon:

$$11 + \quad \begin{array}{c} R^1 \\ R^2 \end{array}\!\!C=N_2 \quad \longrightarrow \quad 407 \quad \xrightarrow{R^1=H} \quad 408 \quad (R^2=H,\text{alkyl})$$

407   408

$$\Big| R^1 = R^2 = CH_3$$

409   410

An analogous diazoketone formation is observed (409) with cyclopropenone and diphenyl ketene[28].

Other 1,3-dipolar reagents show the same mode of reactivity towards cyclopropenones. Thus, the "Munchnones" 412 serving as potential azomethine ylides[259–261] or the nitrile ylids 413[262] effect expansion of the three-membered ring to the 4-pyridone systems $411/414$ as a result of (2 + 3) cycloaddition to the C^1/C^2 bond.

Analogously, the mesoionic N-methyl thiazol-5-ones and 1,3-dithiol-4-ones afforded N-methyl-4-pyridones and thiapyran-4-ones when reacting with diphenyl cyclopropenone and its thione[261]. Benzonitrile oxide apparently gives a 1,3-dipolar cycloaddition to the C=O group of diphenyl cyclopropenone rationalizing the formation of triphenyl-1,3-oxazin-6-one 416[261].

Tetracyano ethylene oxide, however, which represents a potential 1,3-dipole of the carbonyl ylide type, reacts with diphenyl cyclopropenone to give a cycloadduct of probable structure $415/417$[263], which may arise from insertion into the cyclopropenone $C^{1(2)}/C^3$ bond.

In contrast, aziridines (serving as potential azomethine ylides) are capable of different addition modes depending on substitution and reaction conditions. Thus, 3-aroyl-N-alkyl aziridines 418 preferentially add to the carbonyl group of diphenyl cyclopropenone[264, 265] in a (3 + 2) fashion. The resulting oxazolines 419 may act as 1,3-dipoles by adding to the C^1/C^2 bond of diphenyl or cyclohepteno cyclopropenone[266] with concomitant elimination of Schiff base and carbon monoxide leading to furans 420.

79

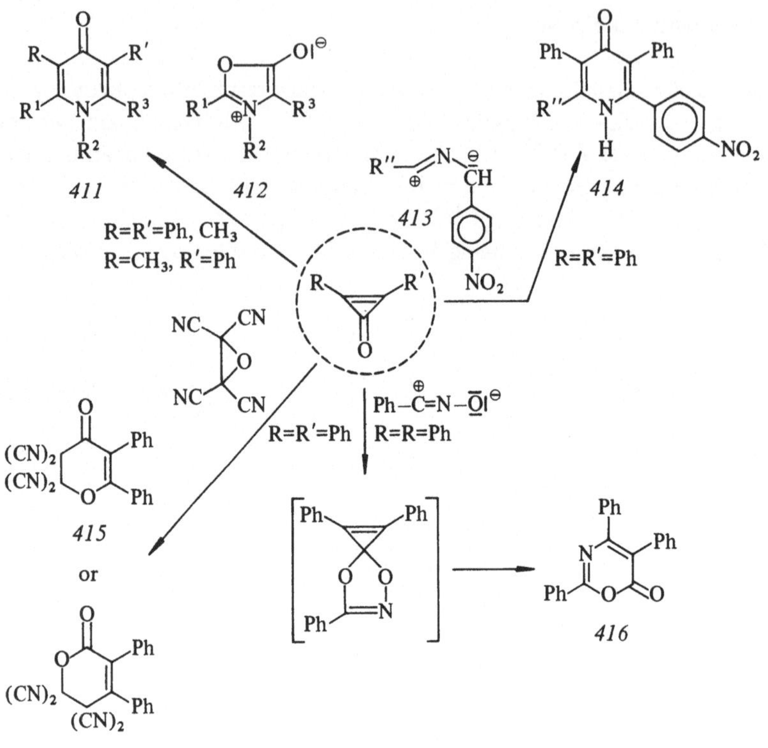
411
412
413
414
R=R'=Ph, CH3
R=CH3, R'=Ph
R=R'=Ph
R=R'=Ph
R=R=Ph
415
or
416

417

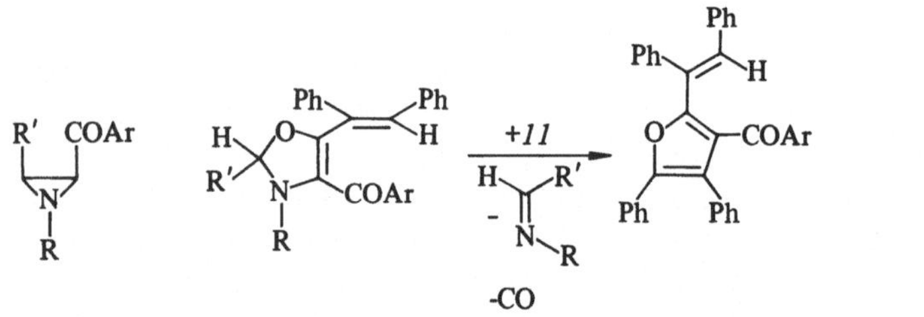
+11
-CO

418 419 420

11

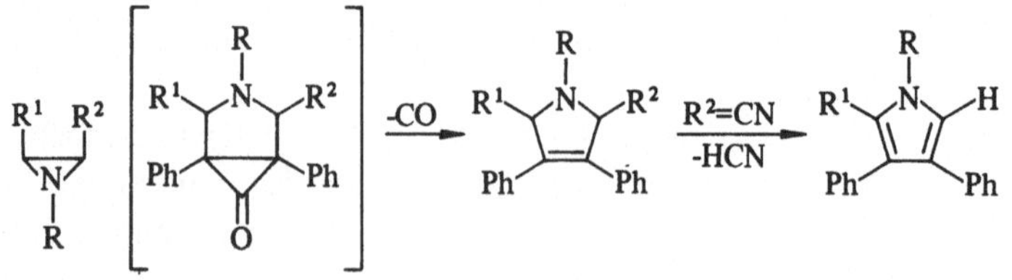
-CO
R²=CN
-HCN

421 422 423

R^2=CO_2R
CN

Carbalkoxy- and cyanosubstituted N-alkyl aziridines 421, however, undergo 1,3-dipolar cycloaddition to the C^1/C^2 bond of diphenyl cyclopropenone followed by elimination of CO to form the dihydro pyrrole derivatives 422, which may lose HCN (when R^2 = CN) yielding pyrroles 423[234]).

The N-trichloroacetyl cyclopropenone imine 424 reacts with 3-aroyl- or 3-carbethoxy-N-alkyl aziridines exclusively across the C=N bond giving rise to spirocyclopropenes 425 and the imidazoline 426[267]).

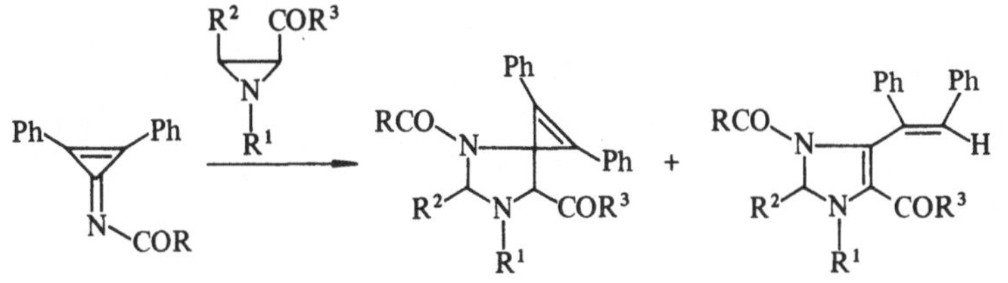

424   425   426

The ring expansion reaction of diaryl cyclopropenones and cyclopropene thiones occurring with pyridinium, sulfonium, and phosphonium enolate betaine 427[268–270) is related to 1,3-dipolar cycloadditions. This process results in formation of 2-pyrones 428 by loss of pyridine (or sulfide or phosphine) and insertion of the remaining fragment C=C–O to the $C^{1(2)}/C^3$ bond of the cyclopropenone:

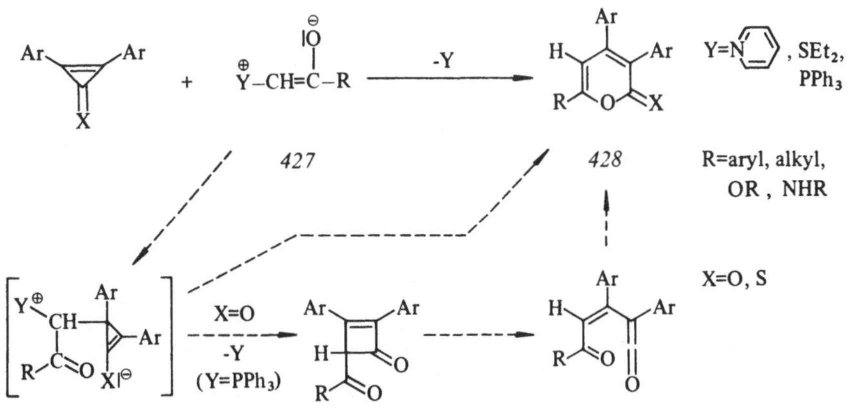

429   430   431

There is mechanistic evidence to show that this formal (3 + 3) cycloaddition starts with attack of betaine-C at $C^{1(2)}$ of the three-ring (429) and leads to 2-pyrone formation either by a "concerted" process (429 ⟶ 428) or stepwise via cyclobutenone and β-acyl vinylketene intermediates (430/431) depending on the "leaving group" Y (as confirmed by results with triafulvenes (see p. 101)). With phosphonium ylides 2-pyrone formation competes with Wittig olefination of the cyclopropenone carbonyl group.

This facile and versatile conversion of cyclopropenones to 2-pyrones is preparatively satisfying (yields are generally up to 80%) and proves to be of general scope;

it could not only be applied to synthesis of "exotic" 2-pyrone systems like *432–434*, but was also found to occur polyfunctionally giving rise to bis- and tris-pyrones *(435/436)*[262, 271]:

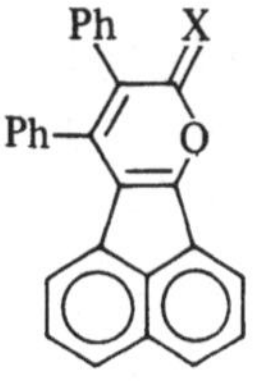

432 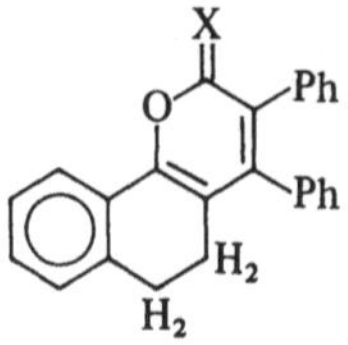   *433* 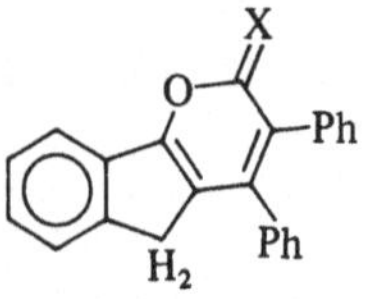   *434*

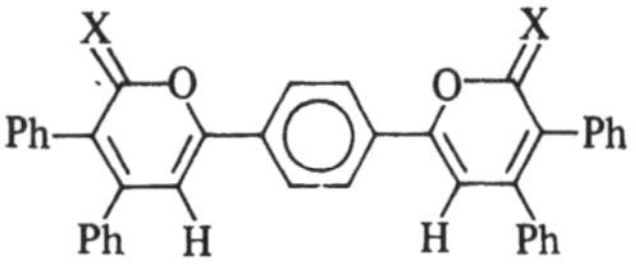

435 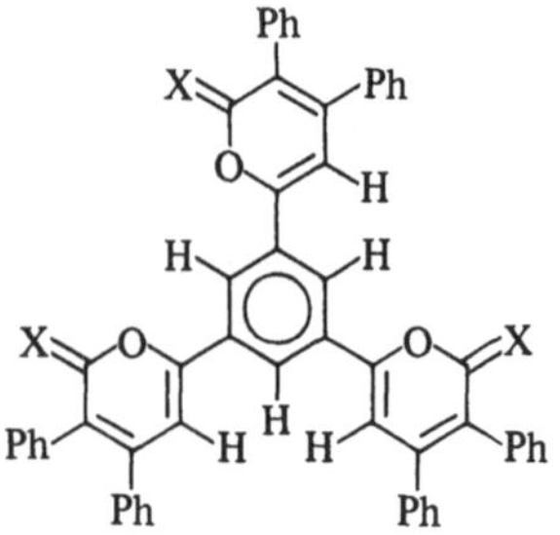   *436*

Interestingly, cyclopropenone exhibited comparable reactivity towards sulfur ylide *437* and phosphorus ylide *439* giving rise to 6-phenyl-2-pyrone and α-naphthol, respectively[197]. Again, the intermediacy of ketenes *438/440* may reasonably explain the formation of these products.

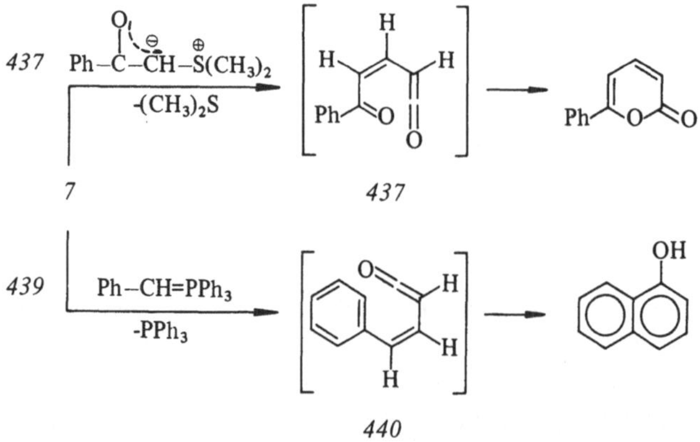

437   *437*

440

Allyl pyridinium betaines *441* isoelectronic with enol betaines *427* likewise reacted with diphenyl cyclopropenone by elimination of pyridine[272, 273]. The product formation, different in aprotic and protic media (phenol *443* in aprotic solvent, $\Delta^{3,5}$-hexadienoic esters *445* in alcohol solvent), suggested that the diene

82

ketene *442* is a central intermediate. It either cyclizes to cyclohexadienone-1,2 *444* (rearranged to phenol *443*) in aprotic medium or may form esters *445* by addition of hydroxylic solvent:

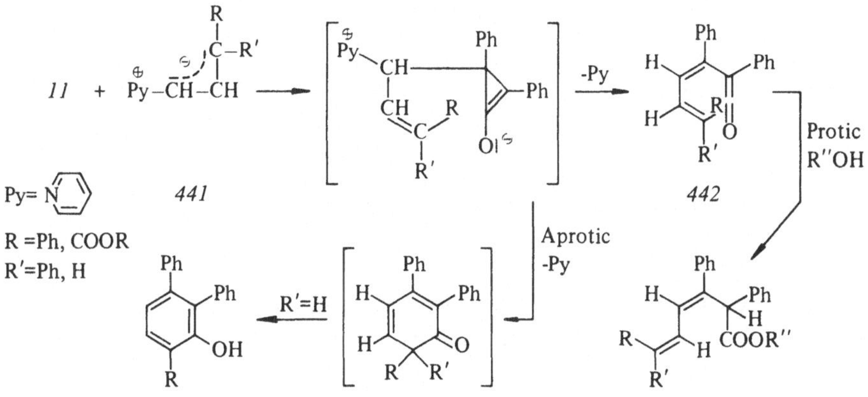

Py= N

441

R = Ph, COOR
R' = Ph, H

443

444

445

442

In agreement with the behavior of ylides *427/441* the pyridinium imine betaines *446* gave rise to the formation of oxazinones *447* on interaction with diphenyl cyclopropenone[274] or its thio analogue[275]:

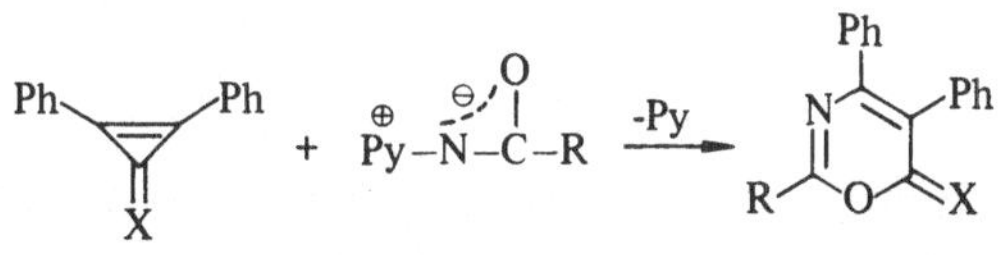

X=O, S   *446*   *447*

The (3 + 3) cycloaddition principle has been extended to the heterocyclic betaines *448* representing aza analogues of ylides *427*. The betaines *448* combined with diphenyl cyclopropenone and its thione[268] to yield the condensed heteroaromatic systems *449*:

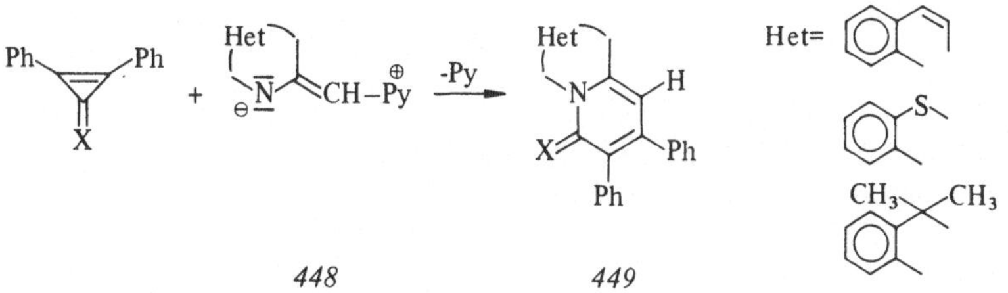

448   *449*

A more complex cycloaddition type is observed when diphenyl cyclopropenone and its thio analogue are reacted with the pyrylium betaine *451*[276] and the products obtained were assigned structures *450* and *452*, respectively.

83

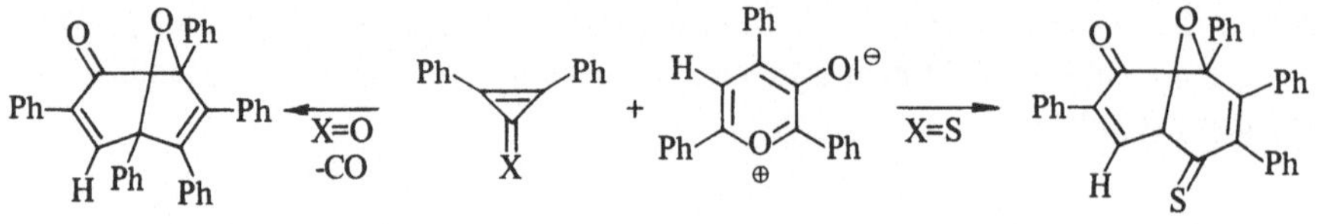

450   451   452

7. Interaction of Cyclopropenones with Transition-metal Compounds

Recent investigations confirmed earlier findings on diphenyl cyclopropenone[277, 278] ruling out the intermediacy of cyclopropenones in the catalytic carbonylation of alkynes:

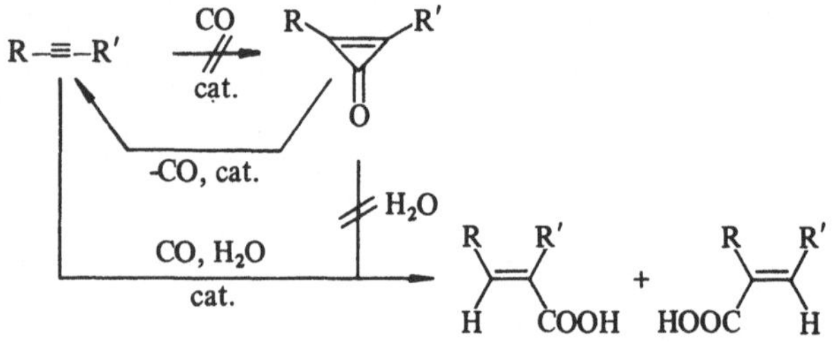

318   319

Thus asymmetric diaryl cyclopropenones were converted to the isomeric acrylic acids *318/319* by aqueous $Ni(CO)_4$ in a similar proportion to that obtained from the corresponding acetylenes by carbonylation with the same catalyst[279], whilst in non-aqueous media carbonyls like $Ni(CO)_4$, $Co_2(CO)_8$, or $Fe_3(CO)_{12}$ effected de-carbonylation[278, 280] probably via metal-complexed intermediates, *e.g.*

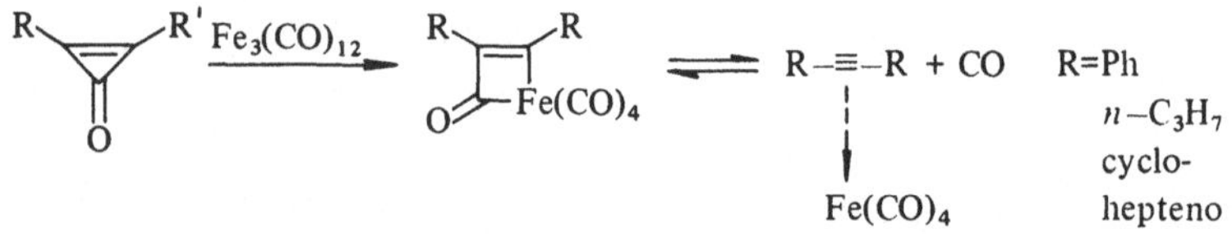

Diphenyl cyclopropenone and $Ni(CO)_4$ were reported[278] to give a transition-metal complex of structure *453*, which was recently re-evaluated in favor of formulation *454* from spectral and chemical evidence[199].

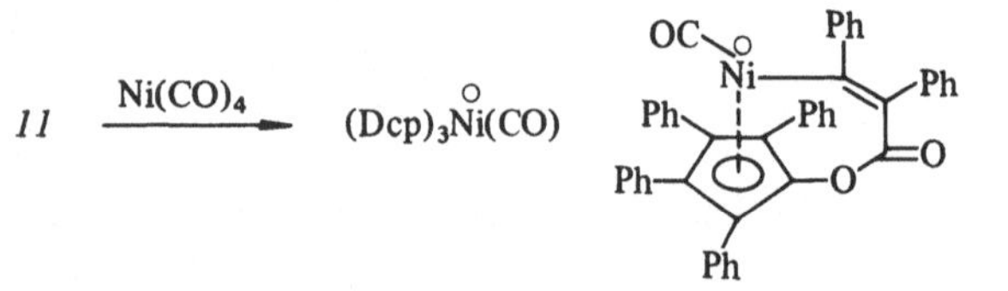

(Dcp)   453   454

A large number of stable planar, tetrahedral, or octahedral complexes of Ib, IIb, and VIIIb elements (Cu^{2+}, Zn^{2+}, Co^{2+}, Ni^{2+}, Ru^{2+}, Rh^{2+}, Pd^{2+}, Pt^{2+}, Pt^{4+}) using cyclopropenones (preferentially the diphenyl compound) as ligands has been evaluated mainly by Bird[281]. Their preparation may start either with metal salts or with carbonyls, as the octahedral Co(II) complex $[Co(Dcp)_6]^{2+}$ may exemplify:

$$Co(ClO_4)_2 + 6 \, Dcp \xrightarrow{\text{EtOH}} [Co(Dcp)_6](ClO_4)_2$$

$$3 \, Co_2(CO)_8 + 12 \, Dcp \xrightarrow{\text{benzene}} 2 \, [Co(Dcp)_6][Co(CO)_4]_2 + 8 \, CO$$

Dcp = diphenyl cyclopropenone

Furthermore, a series of mixed diphenyl cyclopropenone complexes was obtained[282], in which olefins, phosphines, CO, and halide ion served as additional ligands.

In all these complexes the cyclopropenone ligand was shown to interact with the central transition element by means of the carbonyl function from spectroscopic criteria, its donor capacity was compared to pyridine-N-oxide[282].

In contrast, methyl cyclopropenone is reported[283] to react with the Pt-olefin complex 455 at low temperature with replacement of the olefin ligand. In the resulting complex 456 the cyclopropenone interacts with the central atom via the C^1/C^2 "double" bond according to spectroscopic evidence[284]. At elevated temperatures a metal insertion to the $C^{1(2)}/C^3$ bond occurs giving rise to 457. Pt complexes of a similiar type were obtained from dimethyl and diphenyl cyclopropenone on reaction with 455 and their structures were established by X-ray analysis[285].

V. Reactions of Triafulvenes

1. Thermolysis and Photolysis

In contrast to cyclopropenones, most methylene cyclopropene derivatives do not undergo clearly defined thermal transformations; neither cycloreversion analogous to cyclopropenone decarbonylation (to give alkynes and vinylidene carbene or its dimer *459*) nor dimerization giving products of type *458* have been observed[203]. Analogy to cyclopropenone decarbonylation is found, however, on electron impact of triafulvenes (see p. 53).

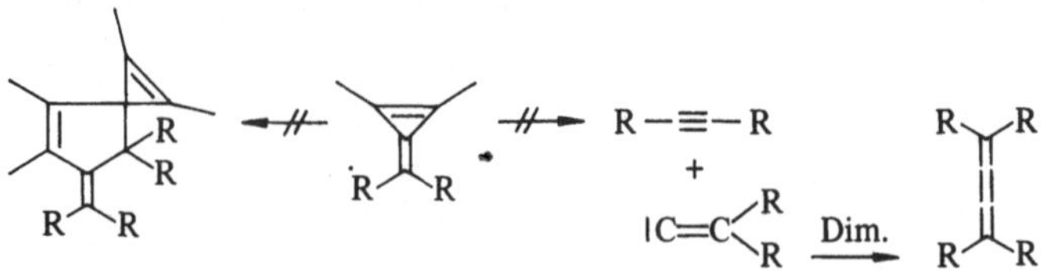

458   459

The photochemical behavior of methylene cyclopropenes is a subject of current investigation[170]. Previous results with some 4,4-diacyl and 4,4-dicyano triafulvenes indicate that mainly dimerization, but sometimes additional solvent incorporation and hydrogen abstraction occurs. In the case of the "photodimer" of 1,2-diphenyl-4,4-diacetyl triafulvene (*180*) the structure *460* can be assigned from spectral evidence:

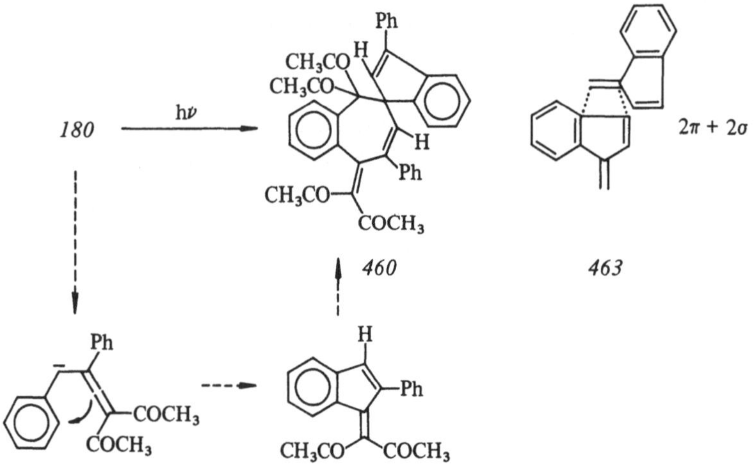

460   463

461   462

Formation of the same dimer from irradiation of benzofulvene *462* suggests that benzofulvene *462* or a photoproduct of it is generated first by photolysis of triafulvene *180* via *461* and then undergoes orbital-symmetry allowed cycloaddition of type *463* to a second benzofulvene molecule.

2. Oxidation and Reduction

1,2-Diphenyl-4,4-dicyano and -4,4-diacetyl triafulvene are remarkably stable against oxidation: with alkaline H_2O_2, which opens the three-ring of diphenyl cyclopropenone to desoxybenzoin (p. 68) no reaction was observed[88]. However, 1,2,3,4-tetra-phenyl-5,6-dimethyl calicene is readily oxidized by atmospheric oxygen[286] giving rise to the allenic ketone 465 probably formed by oxygenation at the triafulvene C^1/C^2 bond via 464. Analogous oxidative cleavage of the three-membered ring is observed with other calicenes[187, 287].

464   465

Accordingly, the cyclopropenylidene anthrones 190/198 were converted by ferric chloride in hydroxylic solvents to the allene ketal 466, whose hydrolysis gives the allenic ketone 467[288]. The dioxolane 468 was obtained from the alkyl-sub-stituted quinocyclopropene 190 in glycol and the ketone 467 in methanol. Apparently FeCl$_3$ served not only as an oxidant, but also as a Lewis acid assisting solvent addition to $C^{1(2)}$ of the triafulvene.

190 : R=n−C$_3$H$_7$
198 : R=Ph   466   467   468

An analogy with reductive dimerization of diphenyl cyclopropenone (p. 58) was found on polarography of 1,2-diphenyl-4,4-dicyano triafulvene (64)[289]. In a one-electron reduction step the cyclopropenyl radical anion 469 is likely to be generated and dimerized to the dianion of tetraphenyl-1,4-dicyanomethyl benzene (470); the dianion 471 could be successively oxidized via the anion radical 472 to the 1,4-quinodimethane derivative 473.

Reductive dimerization of the above type is not observed in the 4,4-diacyl triafulvene series[88]. Instead, 1,2-diphenyl-4,4-diacetyl and -4,4-dibenzoyl triafulvene are readily reduced by means of zinc/acetic acid to "monomeric" products, which are likely to possess structure 474 from their spectral data.

A polarographic study of 4,4-diacyl triafulvenes[290] showed that both oxidative and reductive processes may occur, reduction being somewhat favored over oxidation due to mesomeric effects of the acyl grouping [one-electron reduction: −1.2 to −1.3 V (475); one-electron oxidation: +1.6 to +1.75 V (476)].

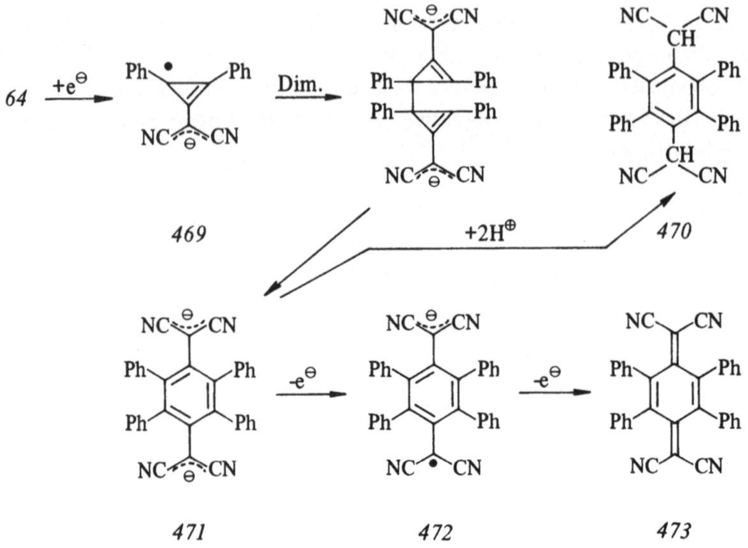

64 $\xrightarrow{+e^{\ominus}}$

469 Dim. → ... $+2H^{\oplus}$ 470

471 $\xrightarrow{-e^{\ominus}}$ 472 $\xrightarrow{-e^{\ominus}}$ 473

$\xrightarrow{\text{Zn/HOAc}}$ R=CH$_3$, Ph

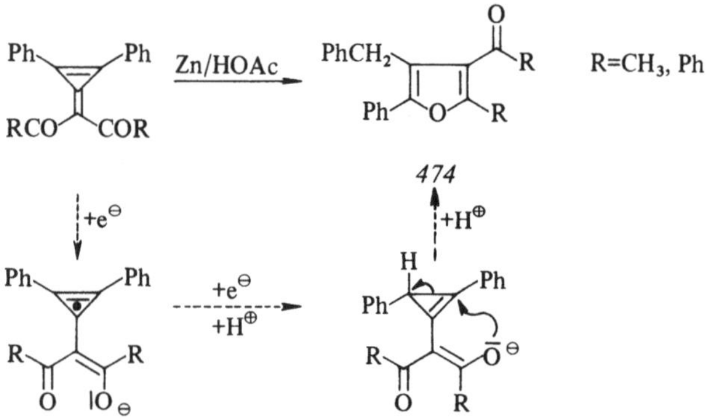

474

$+H^{\oplus}$

$+e^{\ominus}$

$\xrightarrow[+H^{\oplus}]{+e^{\ominus}}$

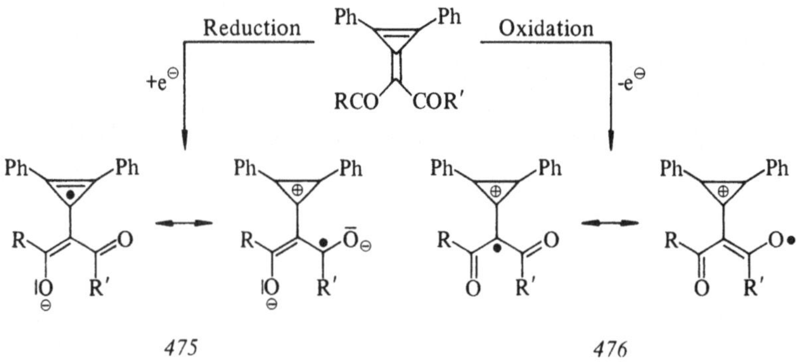

Reduction $\xleftarrow{+e^{\ominus}}$ Oxidation $\xrightarrow{-e^{\ominus}}$

475

476

Interestingly, for the vinylogous triafulvene *96* the reduction potential is drastically lowered to −0.53 V, whilst the oxidation potential stays in the same range as the above examples (+1.58 V). This effect might well reflect the increase of resonance stabilization for the radical anion *477* contributed by the cyano substituents.

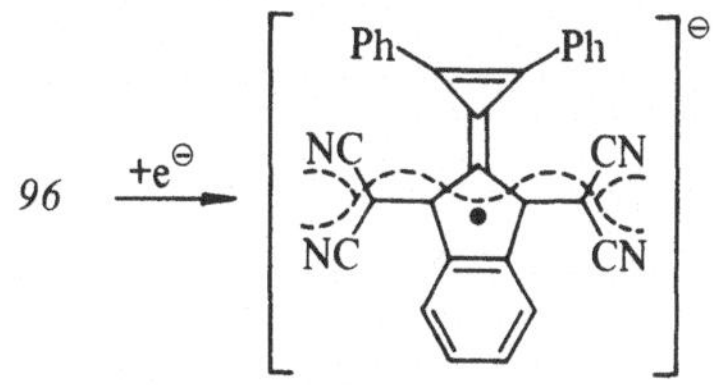

477

The opposite is true for the *o*- and *p*-dicyanomethylene quinocyclopropenes *118–125*[75]. The only electron-transfer observed on polarography corresponds to a one-electron oxidation resulting in the radical cation *479*. A qualitative explanation can be seen in the transformation of the quinocyclopropene into two "Hückel-aromatic" systems favoring oxidation to *479* over reduction to the "antiaromatic" cyclopropenyl radical anion *478*:

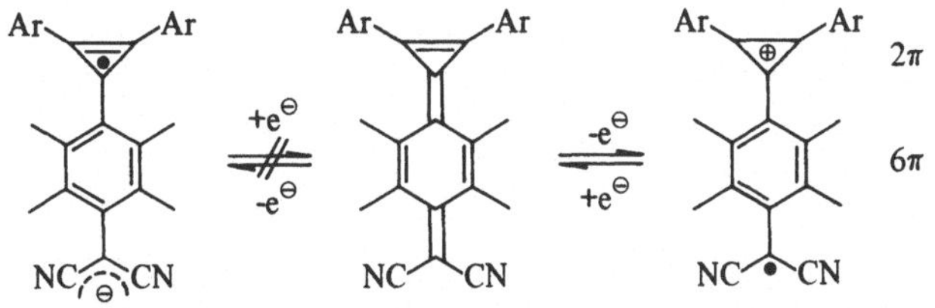

478   *479*

The half-wave redox potentials can be satisfactorily correlated to the energy of the highest occupied molecular orbitals in quinocyclopropenes *118–125* by calculations according to the simple HMO model[75].

1,2,3,4-Tetrachloro-5,6-diphenyl- and -5,6-di-*n*-propyl calicene were found to yield intensely colored radical anions when interacted with alkali metals (Na, K)[291]. As ESR-spectroscopic investigation showed an unpaired electron spin is located on three-ring carbons C^5/C^6 and their substituents.

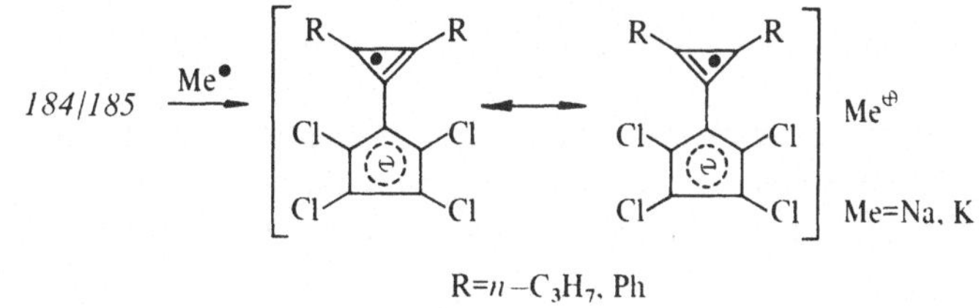

R=*n*−C$_3$H$_7$, Ph

89

3. Reactions with Electrophiles and Nucleophiles

Reactivity of triafulvenes toward electrophiles and nucleophiles is determined by their clearly established (see p. 35) electron distribution. Thus, protonation of the intensely colored triafulvenes $480^{119)}$ and $115^{75)}$ — readily occurring when dissolved in CF_3COOH — leads to colorless cyclopropenium cations $481/482$ by attack of the electrophile at the exocyclic site of the triafulvene system.

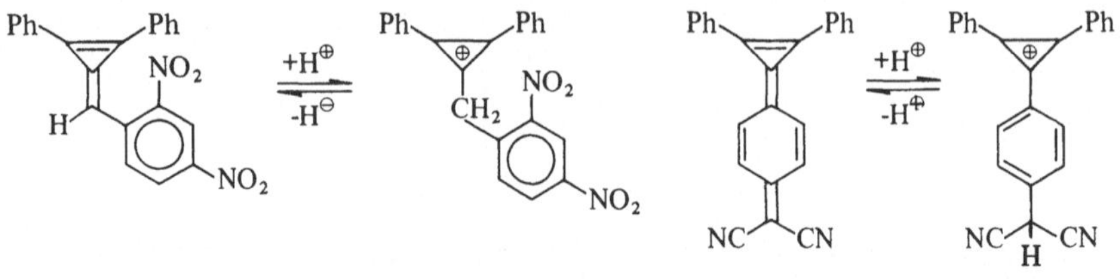

480	481	115	482
(Red)		(Purple)	

  Nucleophiles preferentially interact with triafulvenes by attack at the three-ring carbons, as shown by reactions with water, alcohols, ammonia, and other N- and C-nucleophiles.
  Addition of water to 1,2-diphenyl-4,4-diacetyl triafulvene results in ring-opening to the triketone $483^{88)}$; since in the product the phenyl-bearing centers are separated, the nucleophile must have entered at triafulvene $C^{1(2)}$ rather than at C^3:

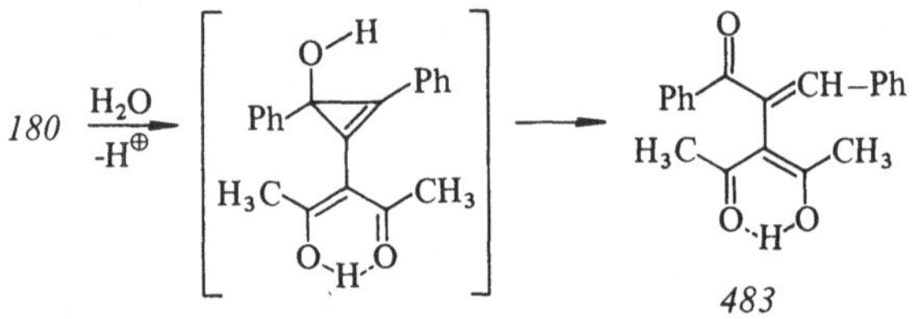

For the azulenic triafulvene 143, however, addition of ethanol is reported[109] to occur at C^3 giving rise to $\Delta^{1,2}$-cyclopropene-3-(azulene-1) ether 484.
  The same direction of nucleophilic attack was found for quinocyclopropene 116 on reaction with Na-ethanolate and Na-acetonitrile, which gave the salts $485/486^{87)}$: As systematic investigations[88] show, the primary attack of N-nucleophiles like ammonia and amines exclusively occurs at $C^{1(2)}$ of the triafulvene system. Further transformation strongly depends on N-substitution and triafulvene type.
  1,2-Diphenyl-4,4-diacetyl triafulvene is converted by ammonia and methyl amine to α-diacetylmethylene azetidine 487, by other primary amines to the conjugated Schiff base 488 and by secondary amines to the cyclic aminal 489. The separation of the former triafulvene C^1/C^2 bond is a common feature of products $488/489$.

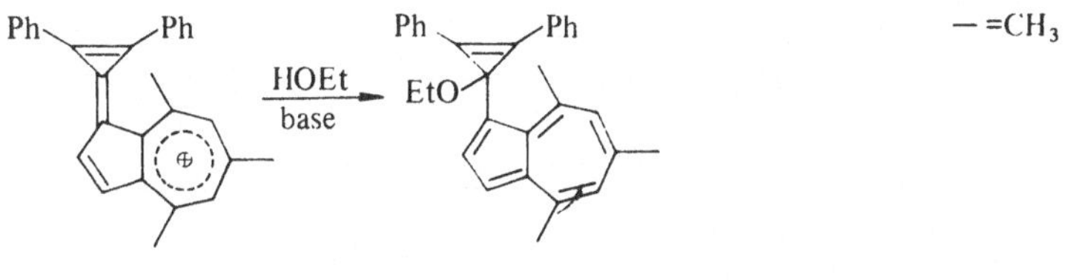

484

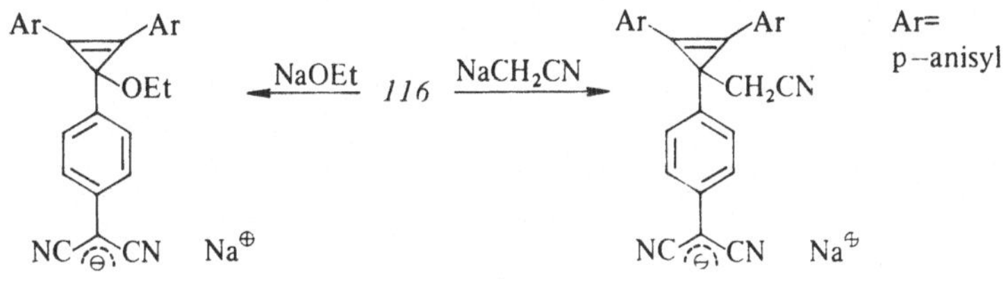

485　　　　　　　　　　*486*

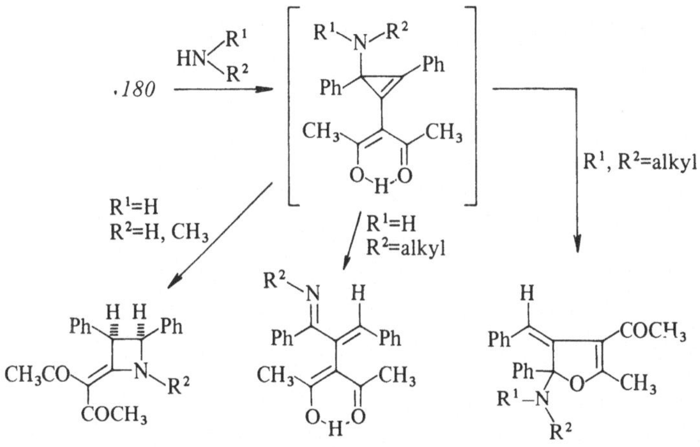

487　　　　　　*488*　　　　　*489*

The cyclic 4,4-diacyl triafulvene *84* shows a reaction pattern with secondary amines different from above; the nucleophile still enters at $C^{1(2)}$ but in subsequent ring-opening the C^1/C^2 bond is maintained and *490* is afforded:

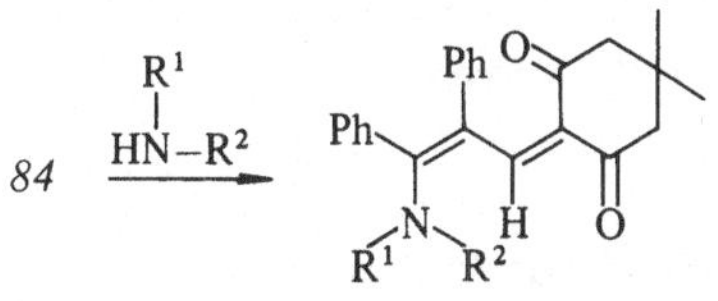

490

1,2-Diphenyl-4,4-dicyano triafulvene reacts readily with two moles of secondary amines building up a highly substituted pyrrole system *491* with incorporation of

91

one of the cyano groups into the heterocycle. The colorless pyrroles *491* are transformed to the purple 3-aza fulvenes *492* by loss of amine on treatment with acids[88].

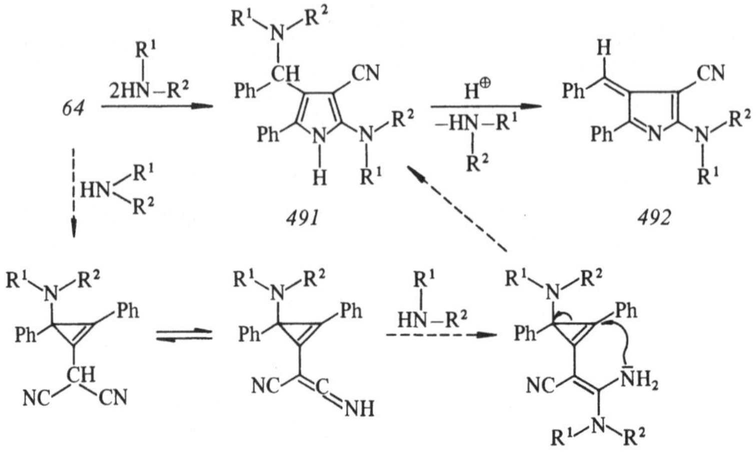

Analogously, 1,2-diphenyl-4,4-diacetyl- and -4,4-dibenzoyl triafulvene are reported[292] to be transformed by hydrazine to pyridazine derivatives *494*, involving attack of the nucleophile at $C^{1(2)}$ and cyclization of intermediate *493*:

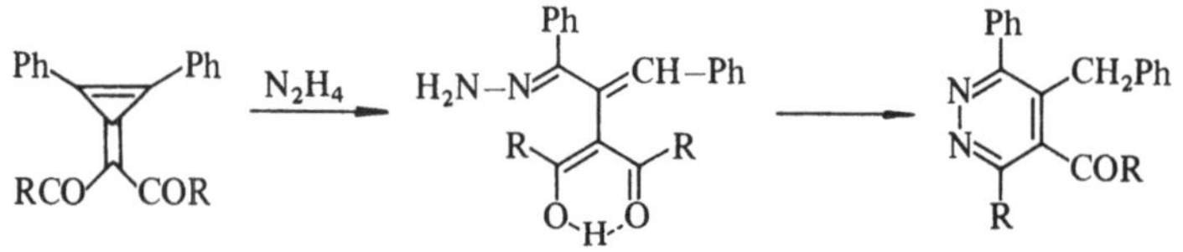

R=CH₃
Ph

4. Reactions with Systems Containing Multiple Bonds

In an attempt toward electron-rich and electron-deficient multiple bonds as well as 1,3-dipoles, the triafulvene system may develop functionalization of a dipolarophilic, dienophilic, and diene component. Rigorous proof for a concerted or a stepwise mechanism, e.g. via dipolar intermediates, for any of the numerous reactions investigated cannot be presented. Therefore the following classification has been chosen from a more or less formal point of view.

a) (2 + 2) and (2 + 4) Cycloaddition Reactions

According to the electron distribution of triafulvenes, systems containing electron-deficient multiple bonds like TCNE, MAA, and ADD interact preferentially with the semicyclic "double" bond of triafulvenes.

Thus TCNE is reported to give with 1,2-diphenyl-4-carbethoxy triafulvene (*69*) the spirocyclohexene *495*[69], the (2 + 2) cycloadduct of the semicyclic methylene cyclopropene bond to TCNE:

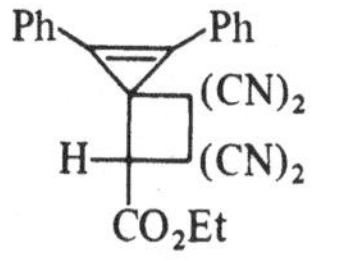

495

In contrast, with the calicene *230* TCNE is attached to five- and three-membered rings in a more complicated cycloaddition mode giving rise to *496*[8]. With a series of other calicenes no cycloaddition, but formation of stable charge-transfer complexes was observed[293].

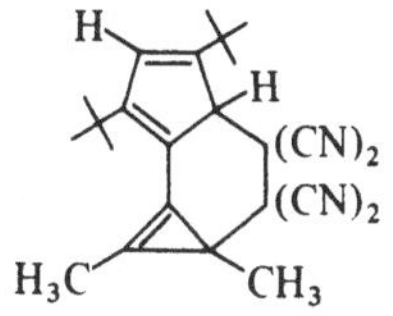

496

Acetylene dicarboxylate and maleic anhydride failed to react with simple methylene cyclopropenes, but reacted readily with calicene derivatives, as shown by Prinzbach[293]. Thus ADD combined with benzocalicene *497* to give the dimethyl triphenylene dicarboxylate *499*, whose formation can be rationalized via (2 + 2) cycloaddition across the semicyclic "double" bond as well as (4 + 2) cycloaddition involving the three-membered ring (*498/501*). The asymmetric substitution of *499* excludes cycloaddition of ADD to the C^1/C^2 triafulvene bond (*500*), which would demand a symmetrical substituent distribution in the final triphenylene derivative.

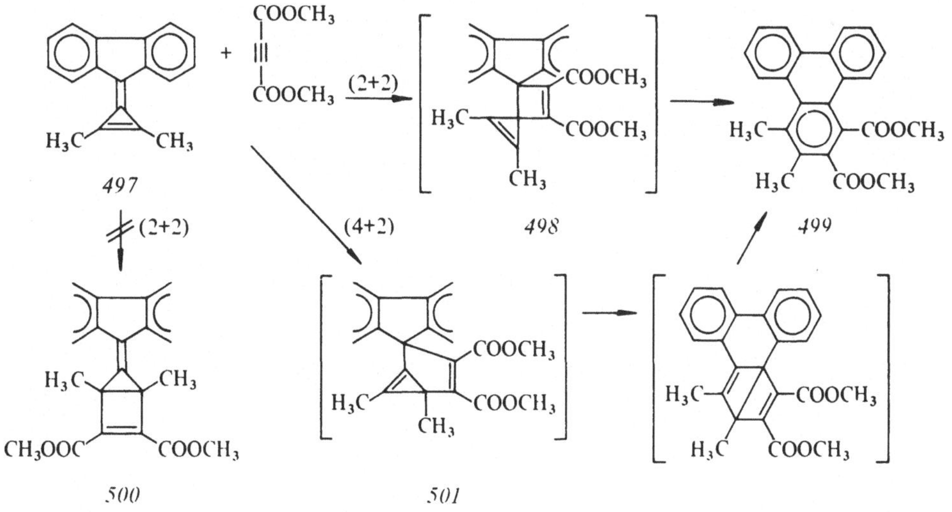

497   *498*   *499*

500   *501*

93

This is in accordance with the findings of Kende[89] who obtained a mixture of isomeric phenanthrene tricarboxylates *503/504* from benzocalicene *502* and ADD.

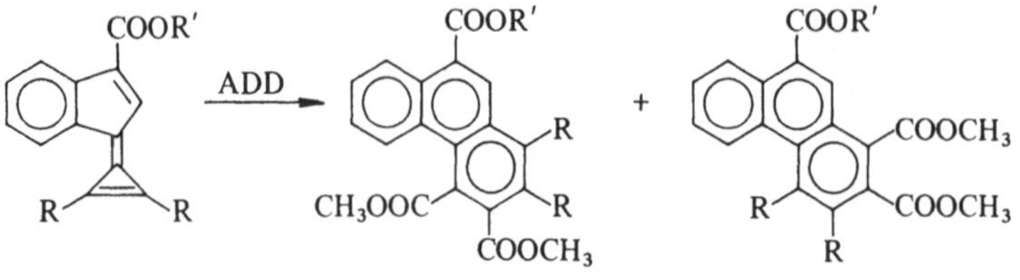

502   *503*   *504*

The addition of maleic anhydride to dibenzo calicene *497*[293] proceeds according to the same (2 + 2) or (4 + 2) mode discussed for the addition of ADD, giving rise to the dihydro triphenylene dicarboxylic anhydride *505*, which is capable of addition of a second molecule of the dienophile (*506*).

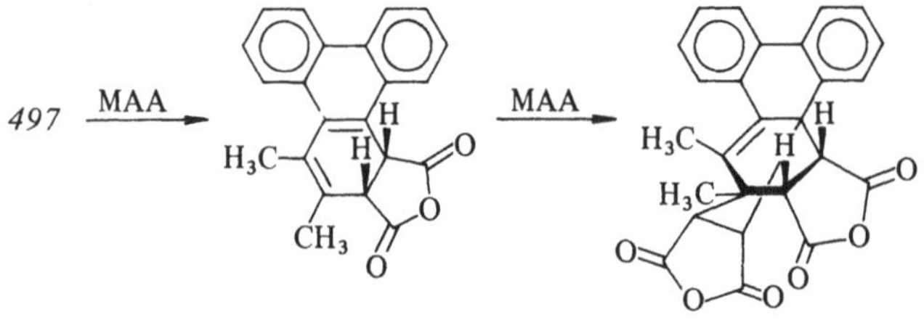

505   *506*

In some cases the C^1/C^2 "double" bond in methylene cyclopropenes and calicenes was found to show dienophilic functionality towards diene components. Thus, diethylamino butadiene combines with *497* to give the Diels-Alder adduct *507*, whose proton-catalyzed elimination of amine interestingly did not lead to the dibenzo heptafulvalene *508*, but to the methylene norcaradiene derivative *509*[293].

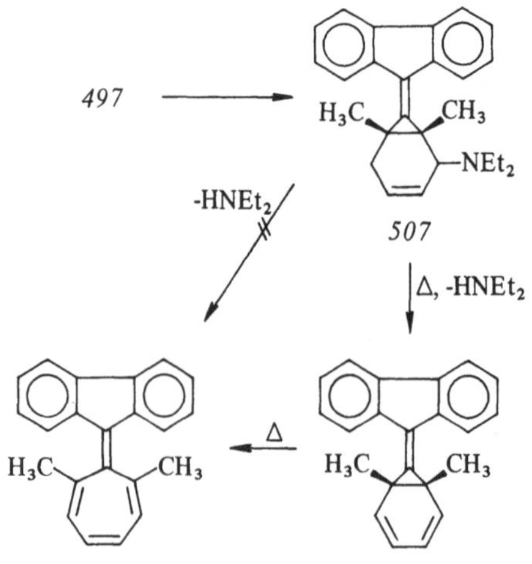

508   *509*

Furthermore, the methyl-substituted triafulvenes *222* underwent Diels-Alder addition to 2-pyrone[55] giving rise to heptafulvenes (*510*) by elimination of carbon dioxide. Extension of this reaction to other triafulvenes was unseccessful.

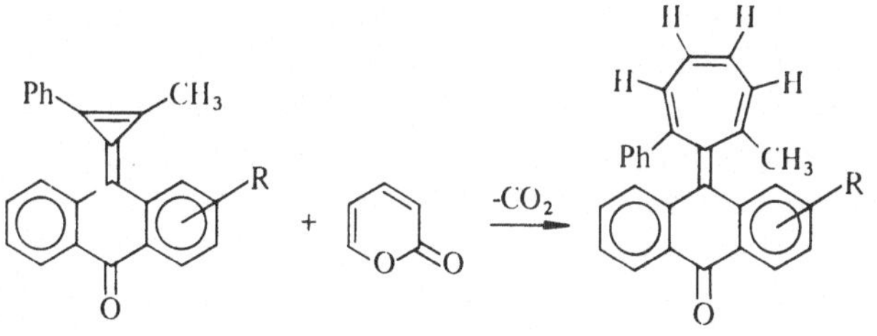

A very elegant formation of heptafulvenes from triafulvenes was found by Gompper[294] utilizing the "push-pull" stabilized cyclobutadiene *511*; in the case of 1,2-diphenyl-4,4-diacetyl triafulvene (*180*) the 4-acyl group caused additional ring closure yielding cyclohepta(b)furan *514*:

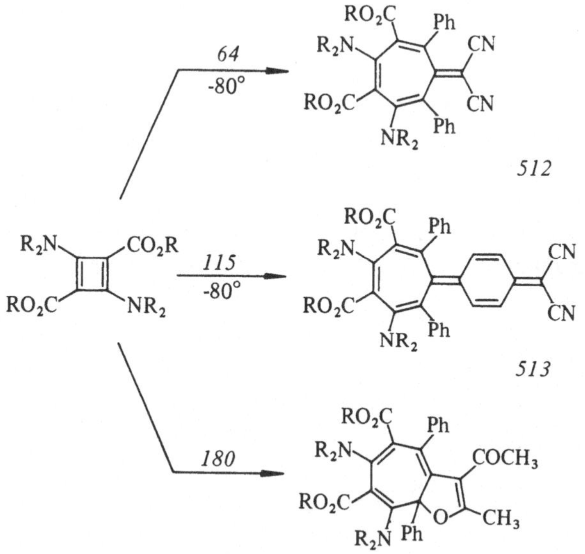

The heptafulvenes *512/513* were believed to originate via dipolar (2 + 2) cycloaddition of a cyclobutadiene to triafulvene C^1/C^2 bond according to the following scheme:

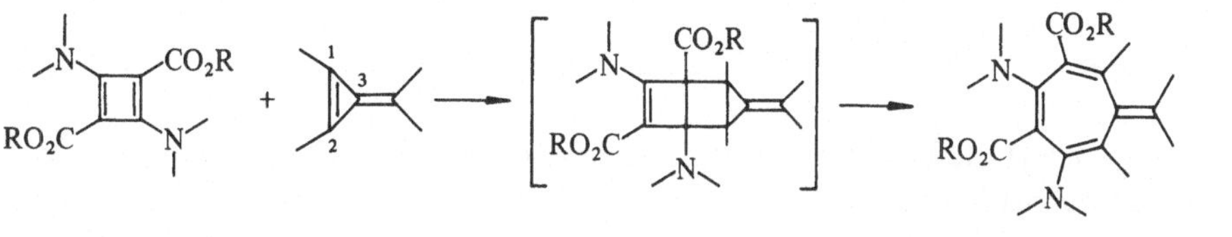

95

b) Reactions with Electron-rich Multiple Bonds (Enamines, Ketene Acetals, and Ynamines)

The rather complex reactivity exhibited by cyclopropenones on interaction with enamines (see p. 74) is not re-found in the reactions of triafulvenes with enamines and ketene acetals. Instead of a $(3 + 3)$ cycloaddition of enamine C=C–N sequence to the $C^{1(2)}/C^3$ bond of triafulvene (as represented by ylide *515*) the addition of the enamine double bond to triafulvene C^1/C^2 bond (operating with cyclopropenones only as a minor side-reaction) predominates in all reactions hitherto investigated. This type of interaction can be envisaged by dipolar intermediates *516* or *517*, whose further conversion is significantly dependent on substitution of triafulvene at the exocyclic carbon.

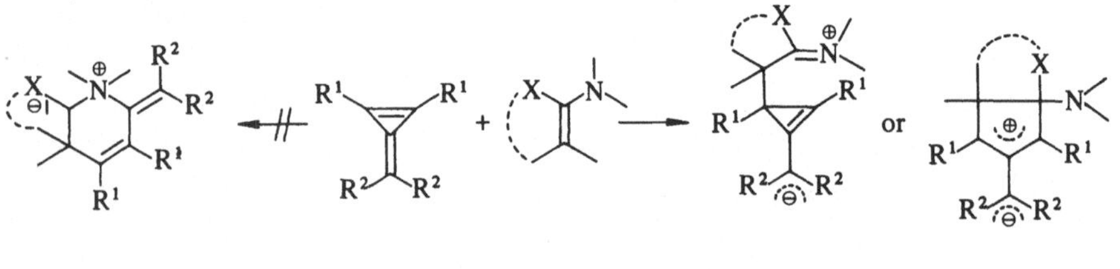

515   *516*   *517*

4,4-Dicyano-substituted triafulvenes react with enamines to produce exclusively the cross-conjugated dicyanomethylene compounds *519*, whose formation can be rationalized by a methylene bicyclo(2,1,0)pentane intermediate *518*[79, 296]. Since cyclanone enamines *520* and other cyclic enamines *522* react analogously, this "C–C-insertion"[237] of the triafulvene ring skeleton into the enamine C=C bond represents a versatile ring expansion mode ($C_n + C_3$), which makes accessible a series of unsaturated medium-ring compounds (*521/523*) that are otherwise difficult to synthesize.

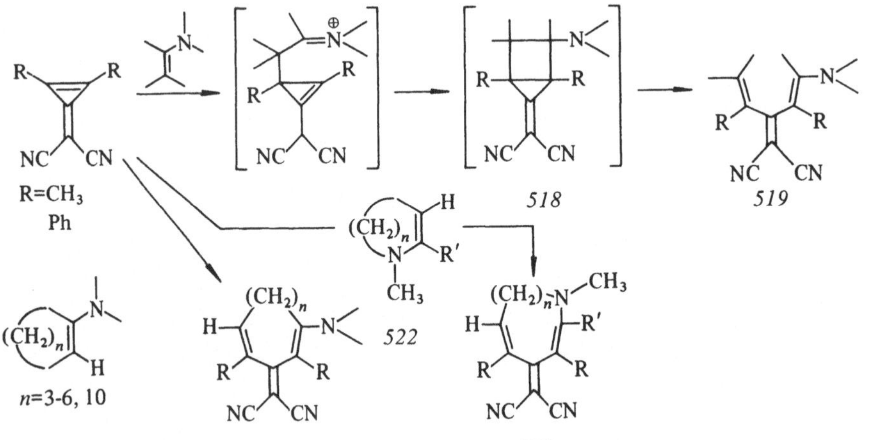

520   *521*   *523*

4,4-Diacyl triafulvenes yield with enamines the 6-amino-5,6-dihydro-6aH-cyclo-penta(b)furans *524* of structure type A/B, type B is only obtained from enamines possessing α-hydrogens[296–298]. The bicyclic species *524* proved to be capable of undergoing some preparatively valuable transformations, as shown by the elimination of amine giving fulvenes of different structure types (*526*/*527*) and the isomerization to cyclopentadiene amines *528* observed in the B series.

525 *526* *527* *528*

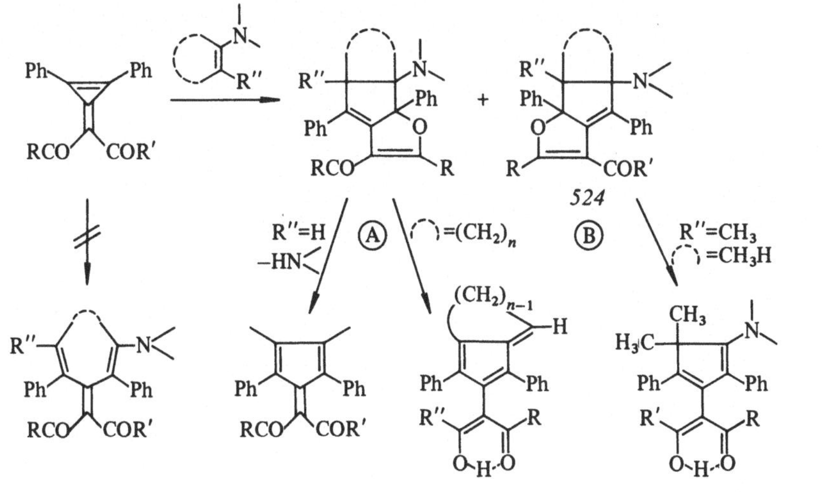

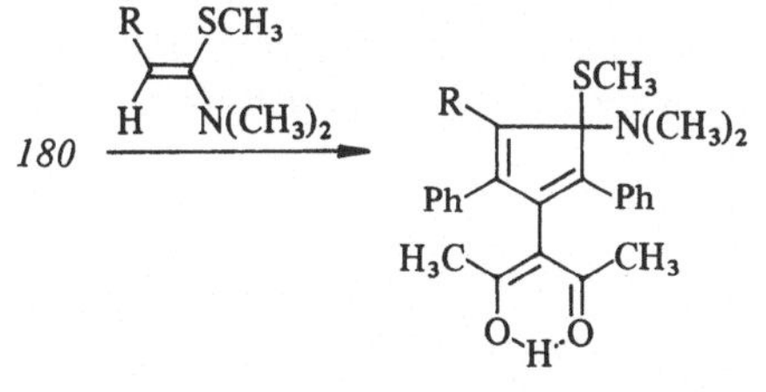

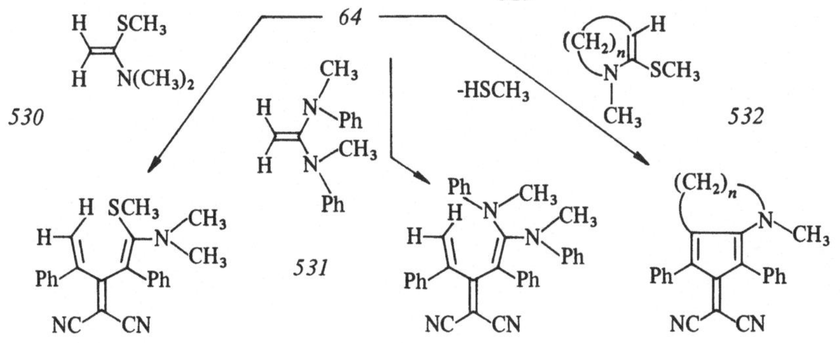

In contrast to the (experimentally well-established) mechanisms of its transformations, the mechanistic aspects concerning formation of *524* are more or less speculative. From the absence of cross-conjugated systems (like *525*) the assumption seems

529

533 *534* *535*

to be justified that dihydro-cyclopenta(b)furans *524* do not arise via a bicyclic inter-mediate (type *518*), but directly via dipoles *516* or *517*.

Ketene acetals show a pattern of product formation very similiar to enamines[79]. Diphenyl-4,4-diacetyl triafulvene is converted to diacetylmethyl cyclopentadiene *529* by S,N-acetals, whilst diphenyl-4,4-dicyano triafulvene undergoes "C–C-inser-tion" to S,N- and N,N-acetals, e.g. *530/531*, resulting in cross-conjugated systems *533/534* by analogy with enamines. Cyclic S,N-acetals *532*, however, yield exclusively the bicyclic fulvenes *535* due to additional loss of methyl mercaptan.

Likewise, Schiff bases of type *379* react with triafulvenes again (see p. 75) via the tautomeric vinylamine form to give products of "C–C-insertion"[253]. In the presence of a COOR group at triafulvene C[4], the cross-conjugated system *536* under-goes facile cyclization to 2-pyridones *537* by loss of alcohol.

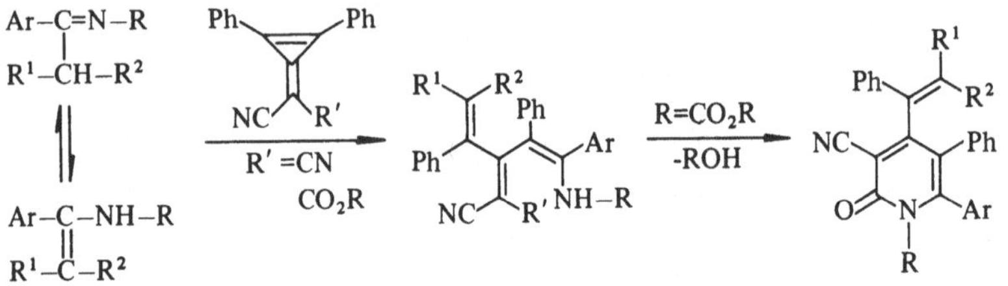

The various transitions of triafulvenes to pentafulvenes achieved by addition of electron-rich double bonds is complemented by the reaction of triafulvenes with ynamines and yndiamines[299], which gives rise to 3-amino fulvenes *539*. This penta-fulvene type deserves some interest for its merocyanine-like "inverse" polarization of the fulvene system and its formation is reasonably rationalized by (2 + 2) cyclo-addition of the electron-rich triple bond to the triafulvene C^1/C^2 bond (probably via the dipolar intermediate *538*):

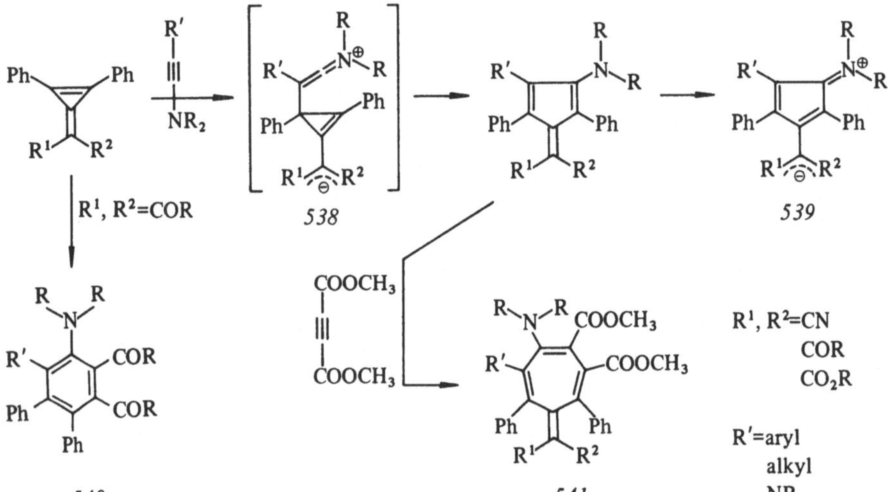

When 4,4-diacyl triafulvenes are reacted, in addition to the deeply colored fulvenes, colorless isomers were also obtained to which structure *540* was assigned.

In contrast to the usual (4 + 2) diene reactivity of fulvenes 3-amino fulvenes *539* in some cases are capable of expanding the five-membered ring to heptafulvenes *541* by addition of acetylene dicarboxylate in a (2 + 2) fashion[299].

The calicene derivative *185* shows ambiguous behavior toward ynamines. Whilst reacting with yndiamine *542* according to the above (2 + 2) mode to give the fulvalene *543*[300], with ynamine *544* a (4 + 2) cycloaddition mode appears to operate which leads to the naphthalene derivative *545*[301]. This is in accordance with the reactivity of other calicenes toward ADD shown earlier (p. 93).

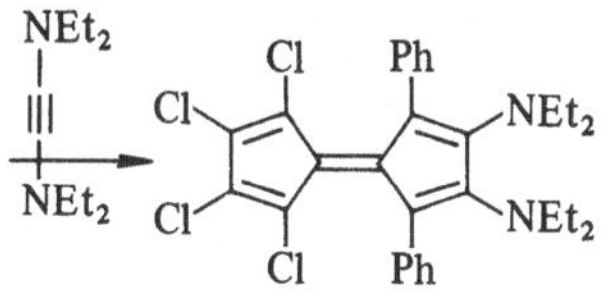

543

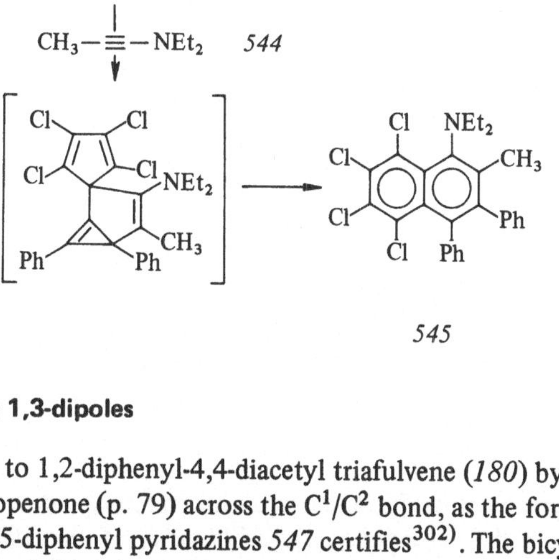

545

5. Reactions with 1,3-dipoles

Diazoalkanes add to 1,2-diphenyl-4,4-diacetyl triafulvene (*180*) by analogy with diphenyl cyclopropenone (p. 79) across the C^1/C^2 bond, as the formation of 4-diacetylmethyl-3,5-diphenyl pyridazines *547* certifies[302]. The bicyclic azo compound *546* was isolated and characterized in the case of R = CH$_3$ and shown to be an intermediate of the diazoalkane reaction by its facile thermal isomerization to pyridazine *547* (R = CH$_3$).

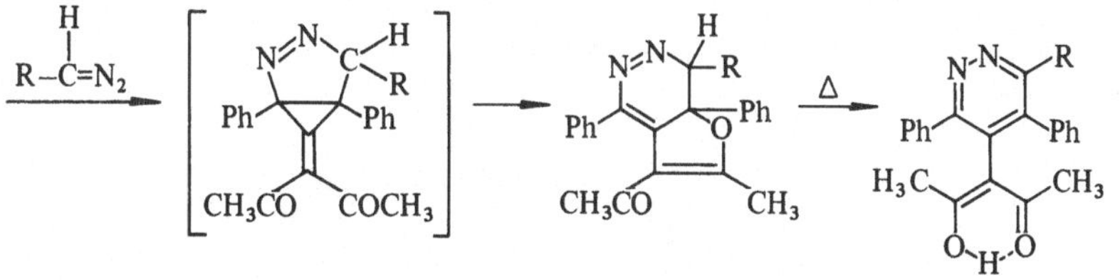

546   *547*

In further agreement with cyclopropenones, "primary" adducts *548* may use a second pattern of stabilization, as observed in the reactions of diazoalkanes with

99

9-(diphenylcyclopropenylidene)anthrone (*198*)[302]. The products are styryl-substituted spiropyrazolenines *550* – E–Z isomeric at the double bond – which thermally either isomerize to the pyrazoles *552* or eliminate nitrogen to give the spirocyclopropenes *551* or undergo acid-catalyzed rearrangement to pyridazines *553*:

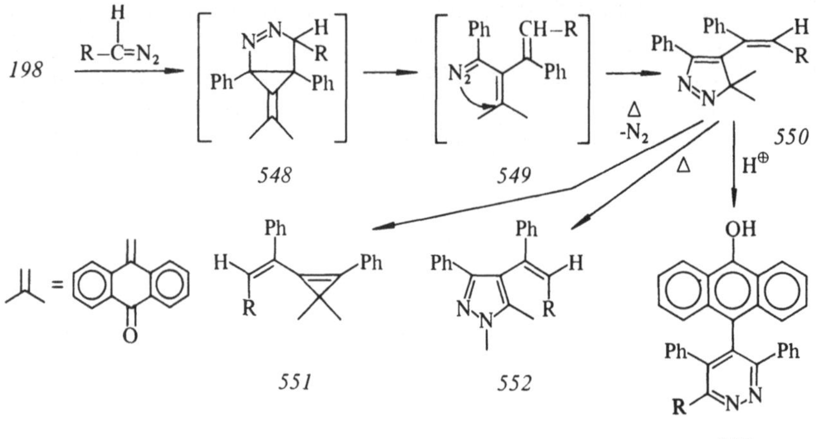

548 *549* *550* *551* *552* *553*

Analogous reactivity was observed by Jones[303] with methylene cyclopropene *554*, which on treatment with diazomethane yielded the pyrazole *556*, thought to arise from thermal isomerization of pyrazolenine *555* formed initially:

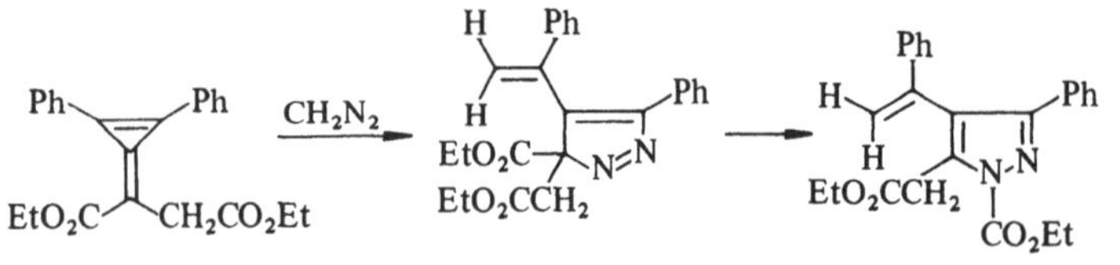

554 *555* *556*

The formation of pyrazolenines demands, as in the cyclopropenone series, the intermediacy of conjugated diazo compounds, *e.g.* *549*, arising from valence tautomerization of diaza bicyclo(3,1,0)hexanes, *e.g.* *548*.

With calicenes, diazoalkanes were found[293] to react in a different manner from other triafulvenes. Thus, dibenzocalicene *497* together with diazomethane gives the product of addition of two moles of the diazo compound *558*, which is likely to arise from primary attack of the 1,3-dipole via (3 + 2) cycloaddition to the triafulvene C^3/C^4 bond (*557*).

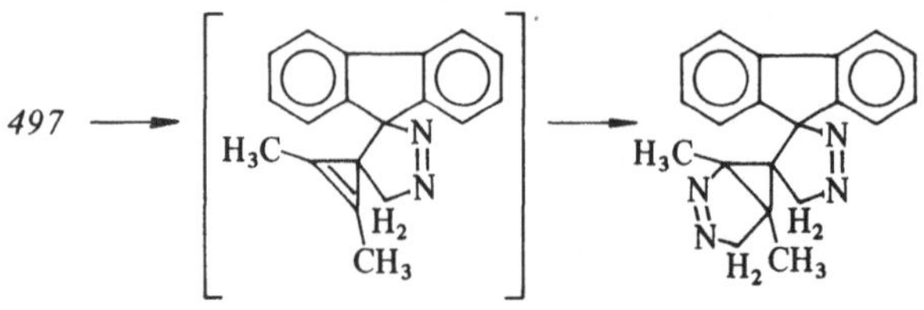

Azomethine ylides such as *412* react with triafulvenes again by analogy with cyclopropenones. (3 + 2) Cycloaddition of the 1,3-dipole to the C^1/C^2 bond and subsequent loss of CO_2 produces 1,4-dihydro-4-methylene-N-alkyl pyridines *559*, which as merocyanines show marked solvatochromic and thermochromic effects[260].

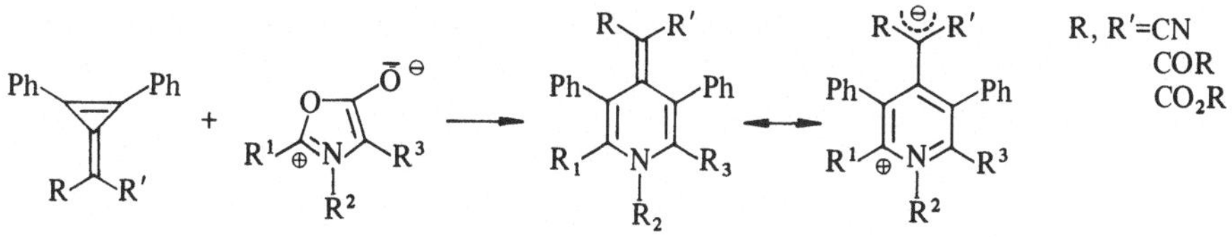

412   *559*

Likewise, pyridinium and sulfonium enolate betaines *427* react with 4,4-diacyl triafulvenes to give ring expansion to the six-membered ring of 2-diacylmethylene pyrane *560*[269]:

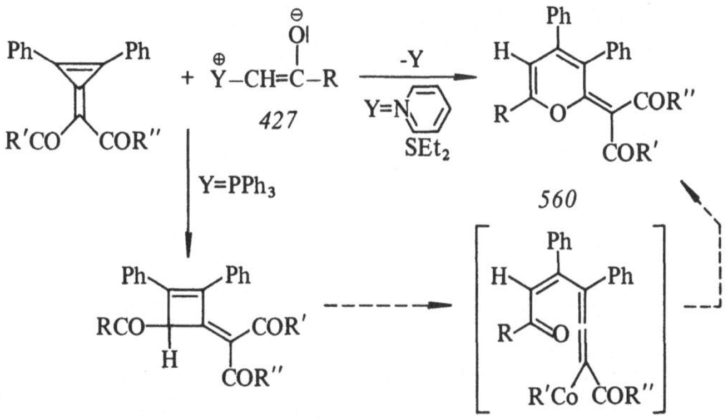

561   *562*

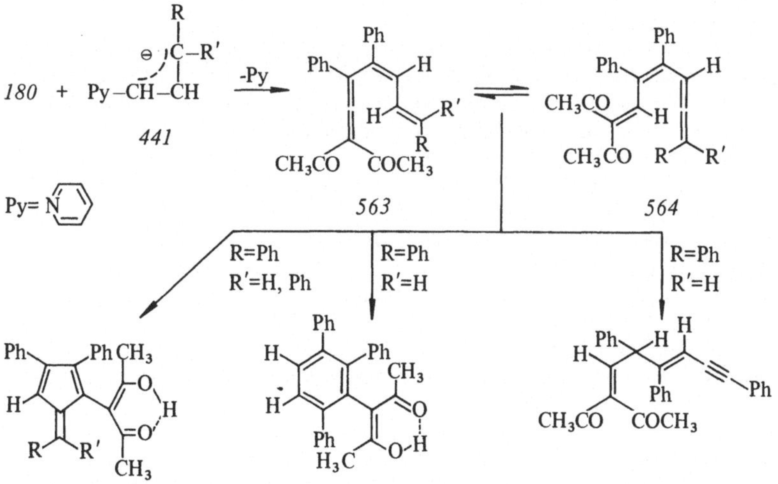

563   *564*

565   *566*   *567*

101

In contrast, the corresponding phosphorus ylides show insertion of the ylidic carbon fragment into the $C^{1(2)}/C^3$ bond of 4,4-diacyl triafulvenes giving rise to α-acyl diacyl-methylene cyclobutenes 561[269], which are isomerized thermally to the 2-methylene pyranes 560, probably via the allene 562.

This unexpected expansion of the triafulvene skeleton to a four-membered ring system presents further evidence in support of the reaction scheme of triafulvenes toward ylides 427 suggested for cyclopropenones (p. 81).

The versatility of this triafulvene reaction type is demonstrated by the interaction of allyl pyridinium betaines 441 and 1,2-diphenyl-4,4-diacetyl triafulvene[272], which gives rise to fulvenes 565, benzene derivatives 566, or acyclic systems 567; these products are likely to result from an allenic precursor 563 and its isomer 564 originating from a 1,5-H-shift.

Acknowledgement. The authors are deeply indebted to Prof. Dr. David St. C. Black, Visiting Professor at the University of Würzburg 1974, Monash University, Clayton (Australia) for his encouraging and consistent help during the preparation of the English manuscript.

Special thanks are due to Prof. Dr. Adolf W. Krebs, University of Heidelberg (BRD) and to Prof. Dr. Herman L. Ammon, University of Maryland, College Park, Maryland (USA) for helpful criticism and valuable suggestions on the structural part of this article.

The assistance of Dipl. Chem. Heinz Ehrhardt during the final preparation of the manuscript is gratefully appreciated.

VI. References and Notes

[1] Lloyd, D., in: Organic chemistry (series 1), aromatic compounds, p. 179. London: Butterworths 1973.

[2] It was suggested by Agranat, I., and Bergmann, E. D.: Tetrahedron Lett. *1966*, 2373, to use the name "triafulvene" for methylene cyclopropenes as well as for cyclopropenones and their functional derivatives. In this article, the term "triafulvene" is restricted to methylene cyclopropene.

[3] For the three-ring carbons a numbering was chosen which can be commonly used for cyclopropenone and triafulvene. Although not completely in accord with strict IUPAC numbering rules[10] our suggestion accomodates "triafulvenes" to the numbering in fulvenes of higher ring size, in which the exocyclic carbon normally bears the highest number, *e.g.* 6 in pentafulvene, 8 in heptafulvene.

[4] Krebs, A. W.: Angew. Chem. 77, 10 (1965); Angew. Chem. Int. Ed. Engl. 4, 19 (1965).

[5] Wendisch, D. in: Methoden der organischen Chemie (Houben-Weyl-Müller) Bd. IV/3, S. 729. Stuttgart: Georg Thieme Verlag 1971.

[6] Lloyd, D. in: Carbocyclic *Non-*benzenoid *A*romatic Compounds, p. 24. Amsterdam-London-New York: Elsevier Publishing Company 1966.

[7] Deem, M. L.: Synthesis *1972*, 675.

[8] Prinzbach, H.: Pure Appl. Chem. 28, 281 (1971) (Calicenes).

[9] Closs, G. L. in: Advances in alicyclic chemistry, Vol. I, p. 114. New York: Academic Press 1966.

[10] Potts, K. T., Baum, J. S.: Chem. Reviews 74, 189 (1974).

[11] Bergmann, E. D.: Chem. Reviews 68, 41 (1968).

[12] Yoshida, Z.: Topics Curr. Chem./Fortschr. Chem. Forsch. 40, 47 (1973) (aminosubstituted cyclopropenones).

[13] Breslow, R.: J. Amer. Chem. Soc. *79*, 5318 (1957).

[14] Baird, N. C., Dewar, M. J. S.: J. Amer. Chem. Soc. *89*, 3966 (1967).

[15] Breslow, R., Haynie, R., Mirra, J.: J. Amer. Chem. Soc. *81*, 247 (1959).

[16] Bird, C. W., Harmer, A. F.: J. Chem. Soc. C *1969*, 959.

[17] Volpin, M. E., Koreshkov, Y. D., Kursanov, D. N.: Izv. Akad. SSSR *1959*, 560.

[18] Dehmlow, E. V.: Angew. Chem. *86*, 187 (1974); Angew. Chem. Int. Ed. Engl. *13*, 170 (1974).

[19] Dehmlow, E. V.: Chem. Ber. *100*, 3829 (1967).

[20] Dehmlow, E. V.: Chem. Ber. *101*, 427 (1968).

[21] Seyferth, D., Damrauer, R.: J. Org. Chem. *31*, 1660 (1966).

[22] Dehmlow, E. V.: J. Organometal. Chem. *6*, 296 (1966).

[23] Breslow, R., Altman, L. J.: J. Amer. Chem. Soc. *88*, 504 (1966).

[24] Crabbé, P., Grezemkovsky, R., Knox, L.: Bull. Soc. Chim. France *1968*, 789.

[25] Anderson, P., Crabbé, P., Cross, A. D., Fried, J. H., Knox, L., Murphy, J., Velarde, E.: J. Amer. Chem. Soc. *90*, 3888 (1968).

[26] Crabbé, P., Carpio, H., Velarde, E., Fried, J. H.: J. Org. Chem. *38*, 1478 (1973).

[27] Breslow, R., Ryan, G.: J. Amer. Chem. Soc. *89*, 3073 (1967).

[28] Breslow, R., Oda, M.: J. Amer. Chem. Soc. *94*, 4787 (1972).

[29] Tobey, S. W., West, R.: J. Amer. Chem. Soc. *88*, 2478 (1966).

[30] Breslow, R.: private communication.

[31] Baucom, K. B., Butler, G. B.: J. Org. Chem. *37*, 1730 (1972).

[32] West, R., Chickos, J., Osawa, E.: J. Amer. Chem. Soc. *90*, 3885 (1968).

[33] Tobey, S. W., West, R.: J. Amer. Chem. Soc. *86*, 4215 (1964).

[34] West, R., Zecher, D. C.: J. Amer. Chem. Soc. *89*, 152 (1967).

[35] West, R., Zecher, D. C., Goyert, W.: J. Amer. Chem. Soc. *92*, 149 (1970).

[36] West, R., Zecher, D. C., Tobey, S. W.: J. Amer. Chem. Soc. *92*, 168 (1970).

[37] Chickos, J. S., Patton, E., West, R.: J. Org. Chem. *39*, 1647 (1974).

[38] Bauer, C. B., LeGoff, E.: Synthesis *1970*, 544.

[39] Agranat, I., Dinur, S.: Chem. Scr. *5*, 137 (1974).

[40] Law, D. C. F., Tobey, S. W., West, R.: J. Org. Chem. *38*, 768 (1973).

[41] Breslow, R., Posner, J., Krebs, A.: J. Amer. Chem. Soc. *85*, 234 (1963).

[42] Breslow, R., Eicher, Th., Krebs, A., Peterson, R. A., Posner, J.: J. Amer. Chem. Soc. *87*, 1320 (1965).

[43] Breslow, R., Altman, L. J., Krebs, A., Mohacsi, E., Murata, I., Peterson, R. A., Posner, J.: J. Amer. Chem. Soc. *87*, 1326 (1965).

[44] Ciabattoni, J., Nathan III, E. C.: J. Amer. Chem. Soc. *91*, 4766 (1969).

[45] Krebs, A., Breckwoldt, J.: Tetrahedron Lett. *1969*, 3797.

[46] Hünig, S., Kiessel, M.: Chem. Ber. *91*, 380 (1958).

[47] Suda, M., Masamune, S.: J. C. S. Chem. Commun. *1974*, 504.

[48] McCorkindale, N. J., Raphael, R. A., Scott, W. T., Zwanenburg, B.: J. C. S. Chem. Commun. *1966*, 133.

[49] Dehmlow, E. V.: Tetrahedron Lett. *1972*, 1271. This reaction principle was adapted for the synthesis of deltic acid (dihydroxycyclopropenone) from bis(trimethylsilyl)squarate: Eggerding, D., West, R., J. Amer. Chem. Soc. *97*, 207 (1975).

[50] Whitman, P. J., Trost, B. M: J. Amer. Chem. Soc. *91*, 7534 (1969); J. Amer. Chem. Soc. *96*, 7421 (1974).

[51] Farnum, D. G., Chickos, J., Thurston, P. E.: J. Amer. Chem. Soc. *88*, 3075 (1966).

[52] Farnum, D. G., Thurston, P. E.: J. Amer. Chem. Soc. *86*, 4206 (1964).

[53] Carter, F. L., Frampton, V. L.: Chem. Reviews *64*, 497 (1964).

[54] Eicher, Th., Pelz, N.: Tetrahedron Lett. *1974*, 1631.

[55] Eicher, Th., Pelz, N.: unpublished results.

[56] Krebs, A.: private communication.

[57] Kerber, R. C., Hsu, C.-M.: J. Amer. Chem. Soc. *95*, 3239 (1973).

[58] Tobey, S. W.: The Jerusalem Symposia on Quantum Chemistry and Biochemistry, Vol. III, p. 351. Jerusalem: The Israel Academy of Sciences and Humanities 1971.

59) Eicher, Th., Krüger, N., Weber, J. L.: unpublished results.
60) Eicher, Th., Hansen, A.-M.: Chem. Ber. *102*, 319 (1969).
61) Eicher, Th., Pelz, N.: Chem. Ber. *103*, 2647 (1969).
62) Dehmlow, E. V.: Chem. Ber. *101*, 410 (1968).
63) Eicher, Th., Koth, D.: unpublished results.
64) Yoshida, Z., Konishi, H., Tawara, Y., Ogoshi, H.: J. Amer. Chem. Soc. *95*, 3043 (1973).
65) Krull, J. S., D'Angelo, P. F., Arnold, D. R., Hedaya, E., Schissel, P. O.: Tetrahedron Lett. *1971*, 771.
66) Agranat, I., Pick, M. R.: Tetrahedron Lett. *1973*, 4079.
67) Andreades, S.: J. Amer. Chem. Soc. *87*, 3941 (1965).
68) Camaggi, G., Gozzo, F.: J. Chem. Soc. C *1970*, 178.
69) Battiste, M. A.: J. Amer. Chem. Soc. *86*, 942 (1964).
70) Krivun, S. V., Semenov, N. S., Baranov, S. N., Dulenko, V. I.: Z. Obsc. Chim. *40*, 1904 (1970).
71) Bergmann, E. D., Agranat, I.: J. Chem. Soc. C *1968*, 1621.
72) Jones, W. M., Pyron, R. S.: Tetrahedron Lett. *1965*, 479.
73) Bergmann, E. D., Agranat, I.: J. Amer. Chem. Soc. *86*, 3587 (1964).
74) Kende, A. S., Izzo, P. T.: J. Amer. Chem. Soc. *86*, 3587 (1964).
75) Eicher, Th., Eiglmeier, K.: Chem. Ber. *104*, 605 (1971).
76) Eicher, Th., Löschner, A.: Z. Naturforsch. *21b*, 295 (1966).
77) Eicher, Th., Pfister, Th., Krüger, N.: Organic Preparations and Procedures International *6*, 63 (1974).
78) Eicher, Th., Löschner, A.: Z. Naturforsch. *21b*, 899 (1966).
79) Eicher, Th., Krüger, N.: unpublished results.
80) Agranat, I., Barak, A., Pick, M. R.: J. Org. Chem. *38*, 3064 (1973).
81) Agranat, I., Loewenstein, R. M. J., Bergmann, E. D.: J. Amer. Chem. Soc. *90*, 3278 (1968).
82) Asao, T., Yagihara, M., Kitahara, Y.: Bull. Chem. Soc. Japan *46*, 1008 (1973).
83) Toda, F.: Chem. Lett. *1972*, 621.
84) Wagner, H.-U., Seidl, R., Fauß, H.: Tetrahedron Lett. *1972*, 3883.
85) Jones, W. M., Denham, J. M.: J. Amer. Chem. Soc. *86*, 944 (1964).
86) Semmelhack, M. F., DeFranco, R. J.: J. Amer. Chem. Soc. *94*. 2116 (1972).
87) Eicher, Th., Eiglmeier, K.: unpublished results.
88) Eicher, Th., Ehrhardt, H.: unpublished results.
89) Kende, A. S., Izzo, P. T.: J. Amer. Chem. Soc. *87*, 4162 (1965).
90) Ueno, M., Murata, I., Kitahara, Y.: Tetrahedron Lett. *1965*, 2967.
91) Prinzbach, H., Seip, D., Fischer, U.: Angew. Chem. *77*, 258 (1965); Angew. Chem. Int. Ed. Engl. *4*, 242 (1965).
92) Jones, W. M., Pyron, R. S.: J. Amer. Chem. Soc. *87*, 1608 (1965).
93) Kende, A. S., Izzo, P. T., MacGregor, P. T.: J. Amer. Chem. Soc. *88*, 3359 (1966).
94) Bergmann, E. D., Agranat, I.: J. C. S. Chem. Commun. *1965*, 512.
95) Prinzbach, H., Woischnik, E.: Helv. Chim. Acta *52*, 2472 (1969).
96) Murata, I., Nakasuji, K, Kume, H.: Chem. Lett. *1973*, 561.
97) Cais, M., Eisenstadt, A.: J. Amer. Chem. Soc. *89*, 5468 (1967).
98) Kende, A. S.: J. Amer. Chem. Soc. *85*, 1882 (1963).
99) Föhlisch, B., Bürgle, P.: Liebigs Ann. Chem. *701*, 67 (1967).
100) Gompper, R., Wagner, H.-U., Kutter, E.: Chem. Ber. *101*, 4123 (1968).
101) Breslow, R., Chang, H. W.: J. Amer. Chem. Soc. *83*, 2367 (1961).
102) Hünig, S.: Angew. Chem. *76*, 400 (1964); Angew. Chem. Int. Ed. Engl. *3*, 548 (1964).
103) Takahashi, K., Takase, K.: Tetrahedron Lett. *1972*, 2227.
104) Takahashi, K., Fujita, T., Takase, K: Tetrahedron Lett. *1971*, 4507.
105) Eicher, Th., Hansen, A.: Tetrahedron Lett. *1967*, 4321.
106) Boyd, G. V., Heatherington, K.: J. C. S. Chem. Commun. *1971*, 346.
107) Boyd, G. V., Heatherington, K.: J. Chem. Soc., Perkin I, *1973*, 2532.
108) Battiste, M. A., Hill, J. H. M.: The Jerusalem Symposia on Quantum Chemistry and Biochemistry, Vol. III, p. 314. Jerusalem: The Israel Academy of Sciences and Humanities 1971.

109) Föhlisch, B., Bürgle, P., Krockenberger, D.: Angew. Chem. 77, 1019 (1965); Angew. Chem. Int. Ed. Engl. 4, 972 (1965).
110) Oda, M., Tamate, K., Kitahara, Y.: J. C. S. Chem. Commun. 1971, 347.
111) Kabuto, C., Oda, M., Kitahara, Y.: Tetrahedron Lett. 1972, 4851.
112) Battiste, M. A., Hill, J. H. M.: Tetrahedron Lett. 1968, 5537; Tetrahedron Lett. 1968, 5541.
113) Lloyd, D., Preston, N. W.: Chem. and Ind. 1969, 1055.
114) Gompper, R., Weiß, R.: Angew. Chem. 80, 277 (1968); Angew. Chem. Int. Ed. Engl. 7, 296 (1968).
115) Eicher, Th., Frenzel, G.: Z. Naturforsch. 20b, 274 (1965).
116) Paquette, L. A., Barton, Th. J., Horton, N.: Tetrahedron Lett. 1967, 5039.
117) Gompper, R., Studeneer, A., Elser, W.: Tetrahedron Lett. 1968, 1019.
118) Krebs, A., Kimling, H.: Angew. Chem. 83, 401 (1971); Angew. Chem. Int. Ed. Engl. 10, 409 (1971).
119) Eicher, Th.: Habilitationsschrift Würzburg 1967.
120) Laban, G., Fabian, J., Mayer, R.: Z. Chem. 8, 414 (1968).
121) Kitahara, Y., Funamizu, M.: Bull. Chem. Soc. Japan 37, 1897 (1964).
122) Föhlisch, B.: private communication cited in Ref.[4].
123) Metzner, P., Vialle, J.: Bull. Soc. Chim. France 1972, 3138.
124) Wohlgemuth, L. G.: Diss. Abstracts 25, 3278 (1964).
125) Jones, W. M., Stowe, M. E., Wells, Jr., E. E., Lester, E. W.: J. Amer. Chem. Soc. 90, 1849 (1968).
126) Halton, B.: Chem. Reviews 73, 113 (1973).
127) Adamson, J., Forster, D. L., Gilchrist, T. L., Rees, C. W.: J. C. S. Chem. Commun. 1969, 221.
128) Ao, M. S., Burgess, E. M., Schauer, A., Taylor, E. A.: J. C. S. Chem. Commun. 1969, 220.
129) Gilchrist, T. L., Rees, C. W.: J. Chem. Soc. C 1969, 1763.
130a) Halton, B., Woolhouse, A. D., Hugel, H. M., Kelly, D. P.: J. C. S. Chem. Commun. 1974, 247.
130b) Chapman, O. L., Mattes, K., McIntosh, C. L., Pacansky, J., Calder, G. V., Orr, G.: J. Amer. Chem. Soc. 95, 6134 (1973).
131) Holmes, J. D., Pettit, R.: J. Amer. Chem. Soc. 85, 2531 (1963).
132) Breslow, R., Peterson, R.: J. Amer. Chem. Soc. 82, 4426 (1960).
133) Benson, R. C., Flygare, W. H., Oda, M., Breslow, R.: J. Amer. Chem. Soc. 95, 2772 (1973).
134) Clark, D. T., Lilley, D. M.: J. C. S. Chem. Commun. 1970, 147.
135) Ammon, H. L.: J. Amer. Chem. Soc. 95, 7093 (1973).
136) Bertelli, D. J., Andrews, Jr., Th. G., Crews, Ph. O.: J. Amer. Chem. Soc. 91, 5286 (1969).
137a) Skancke, A.: Acta. Chem. Scand. 27, 3243 (1973).
137b) Watanabe, A., Yamaguchi, H., Amako, Y., Azumi, H.: Bull. Chem. Soc. Japan 41, 2196 (1968).
138a) Yoshida, Z., Miyahara, H.: Bull. Chem. Soc. Japan 45, 1919 (1972).
138b) See also Andes Hess, Jr., B., Schaad, L. J., Holyoke, Jr., C. W.: Tetrahedron 28, 5299 (1972).
139) Kitahara, Y., Murata, I., Ueno, M., Sato, K.: J. C. S. Chem. Commun. 1966, 180.
140) Murata, I., Ueno, M., Kitahara, Y., Watanabe, H.: Tetrahedron Lett. 1966, 1831.
141) Bergmann, E. D., Agranat, I.: Tetrahedron 22, 1275 (1966).
142) Bentley, J. B., Everard, K. B., Marsden, R. J. B., Sutton, L. E.: J. Chem. Soc. 1949, 2957.
143) Krebs, A., Schrader, B.: Liebigs Ann. Chem. 709, 46 (1967).
144) Pochan, J. M., Baldwin, J. E., Flygare, W. H.: J. Amer. Chem. Soc. 90, 1072 (1968).
145) Benson, R. C., Norris, C. L., Flygare, W. H., Beak, P.: J. Amer. Chem. Soc. 93, 5592 (1971).
146) Breslow, R., Höver, H., Chang, H. W.: J. Amer. Chem. Soc. 84, 3168 (1962).
147) Olah, G. A., Mateescu, G. D.: J. Amer. Chem. Soc. 92, 1430 (1970).
148) Tsukada, H., Shimanouchi, H., Sasada, Y.: Chem. Lett. 1974, 639. The structural data reported in this paper for the hydrate of diphenyl cyclopropenone deviate markedly from those in Ref.[149].
149) Tsukada, H., Shimanouchi, H., Sasada, Y.: Tetrahedron Lett. 1973, 2455.
150) Reed, L. L., Schaefer, J. P.: J. C. S. Chem. Commun. 1972, 528.

151a) Ammon, H. L., Sherrer, C., Eicher, Th., Krüger, N.: unpublished results.

151b) Ammon, H. L., Sherrer, C., Agranat, I.: unpublished results.

152) Kennard, O., Kerr, K. A., Watson, D. G. Fawcett, J. K.: Proc. Roy. Soc. London A. *316*, 551 (1970).

153) Shimanouchi, H., Sasada, Y., Ashida, T., Kakudo, M., Murata, I., Kitahara, Y.: Acta Cryst. B *25*, 1890 (1969).

154) An X-ray analysis of bis(p-chlorophenyl) cyclopropenone has been reported by Peters, K., and Schnering, v. H. G.: Chem. Ber. *106*, 935 (1973); the results, however, are not in accordance with the data found for diphenyl cyclopropenone (see Table 7 and Ref.[135]).

155) Pochan, J. M., Baldwin, J. E., Flygare, W. H.: J. Amer. Chem. Soc. *91*, 1896 (1969).

156) Dietrich, H., Dierks, H.: Angew. Chem. *80*, 487 (1968); Angew. Chem. Int. Ed. Engl. 7, 465 (1968).

157) Trinajstić, N.: Tetrahedron Lett. *1968*, 1529.

158) Sundaralingam, M., Jensen, L. H.: J. Amer. Chem. Soc. *88*, 198 (1966).

159) Ku, A. T., Sundaralingam, M.: J. Amer. Chem. Soc. *94*, 1688 (1972).

160) It is interesting to note that according to Ref.[136] regardless of the model used, the CNDO/2 treatment predicts a larger π-polarization for cyclopropenone than for tropone. This means that the electrostatic work to achieve a cyclopropenium oxide structure is considerably less than for cycloheptatrienylium oxide. From this reason, cyclopropenone seems to be a closer approximation to an "aromatic" system than tropone which can be described better as a polyolefin.

161) Heilbronner, E., Bock, H.: Das HMO-Modell und seine Anwendung – Tabellen berechneter und experimenteller Größen, p. 131. Weinheim: Verlag Chemie 1971.

162) Harshbarger, W. R., Kuebler, N. A., Robin, M. B.: J. Chem. Phys. *60*, 345 (1974).

163a) Schäfer, W., Schweig, A., Maier, G., Sayrac, T., Crandall, J. K.: Tetrahedron Lett. *1974*, 1213.

163b) Kobayashi, T., Nagakura, S., Yoshida, Z., Konishi, H., Ogoshi, H.: Chem. Lett. *1974*, 445.

164) Osawa, E., Kitamura, K., Yoshida, Z.: J. Amer. Chem. Soc. *89*, 3814 (1967).

165) Toda, F., Akagi, K.: Tetrahedron Lett. *1968*, 3735.

166) Krebs, A., Schrader, B.: Z. Naturforsch. *21b*, 194 (1966).

167) Krebs, A., Schrader, B., Höfler, F.: Tetrahedron Lett. *1968*, 5935.

168) Höfler, F., Schrader, B., Krebs, A.: Z. Naturforsch. *24a*, 1617 (1969).

169) Schubert, R., Ansmann, A., Blechmann, P., Schrader, B.: to be published in J. Mol. Structure 1975.
We are indebted to Prof. Dr. B. Schrader for submitting his paper to us prior to publication.

170) Eicher, Th., Graf, R.: unpublished results.

171) Breslow, R., Lockhart, J., Chang, H. W.: J. Amer. Chem. Soc. *83*, 2375 (1961).

172) Julg, A.: J. Chim. physique physico-chim. biol. *50*, 652 (1953).

173) Breslow, R., Groves, J. T.: J. Amer. Chem. Soc. *92*, 984 (1970).

174) Closs, G. L., Closs, L. E., Böll, W. A.: J. Amer. Chem. Soc. *85*, 3796 (1963).

175) Breslow, R., Ryan, G., Groves, J. T.: J. Amer. Chem. Soc. *92*, 988 (1970).

176) Ciganek, E.: J. Amer. Chem. Soc. *88*, 1979 (1966).

177a) Closs, G. L.: private communication cited in Ref.[23].

177b) Closs, G. L., Böll, W. A., Heyn, H., Dev, V.: J. Amer. Chem. Soc. *90*, 173 (1968).

177c) Farnum, D. G., Mehta, G., Silberman, G.: J. Amer. Chem. Soc. *89*, 5048 (1967).

178) Vidal, M., Vincens, M., Arnaud, P.: Bull. Soc. Chim. France *1972*, 665.

179) This has been done for triafulvene *193* in Ref.[74].

180) The pK_a-data were taken from Ebel, H. F.: Die Acidität der CH-Säuren. Stuttgart: Georg Thieme Verlag 1969.

181) Dayal, S. K., Ehrenson, S., Taft, R. W.: J. Amer. Chem. Soc. *94*, 9113 (1972).

182) Agranat, I.: private communication cited in Ref.[135].

183) Pethrick, R. A., Wyn-Jones, E.: Quart. Rev. (London) *23*, 301 (1969).

184) Gleicher, G. J., Arnold, J. C.: Tetrahedron *29*, 513 (1973).

185) Dewar, M. J. S., Kohn, M. C.: J. Amer. Chem. Soc. *94*, 2699 (1972).

186) Kende, A. S., Izzo, P. T., Fulmor, W.: Tetrahedron Lett. *1966*, 3697.

187) Prinzbach, H., Knöfel, H., Woischnik, E.: The Jerusalem Symposium on Quantum Chemistry and Biochemistry, Vol. III, p. 269. Jerusalem: The Israel Academy of Sciences and Humanities 1971.
188) Laban, G., Mayer, R.: Z. Chem. *12*, 20 (1972).
189) Köppel, C., Schwarz, H., Bohlmann, F.: Org. Mass Spectrom. *9*, 324 (1974).
190) Agranat, I.: Org. Mass Spectrom. *7*, 907 (1973).
191) The decarbonylation reactions are extensively reviewed in Ref.[10].
192) For a review see Hoffmann, R. W.: Dehydrobenzene and cycloalkynes. Weinheim: Verlag Chemie 1967.
193) Zecher, D. C., West, R.: J. Amer. Chem. Soc. *89*, 153 (1967).
194) Ciabattoni, J., Berchtold, G. A.: J. Org. Chem. *31*, 1336 (1966).
195) Two additional dimers of diphenyl cyclopropenone of yet unknown structure have been reported: Grigg, R., Jackson, J. L.: J. Chem. Soc. C *1970*, 552.
196) Phenyl methyl cyclopropenone[55] and phenyl methoxy cyclopropenone[51] gave dimers analogous to spirolactone *257*, but the arrangement of substituents is uncertain.
197) Breslow, R., Oda, M., Pecoraro, J.: Tetrahedron Lett. *1972*, 4415.
198) Noyori, R., Umeda, I., Takaya, H.: Chem. Lett. *1972*, 1189.
199) Bird, C. W., Briggs, E. M.: J. Organomet. Chem. *69*, 311 (1974).
200) Toshima, N., Moritani, I., Nishida, S.: Bull. Chem. Soc. Japan *40*, 1245 (1967); Harrison, Jr., E. A.: J. C. S. Chem. Commun. *1970*, 982.
201) Oda, M., Breslow, R., Pecoraro, J.: Tetrahedron Lett. *1972*, 4419.
202) Eicher, Th., Ehrhardt, H., Pelz, N.: Tetrahedron Lett. 1973, 4353.
203) Eicher, Th.: unpublished results.
204) Obata, N., Hamada, A., Takizawa, T.: Tetrahedron Lett. *1969*, 3917.
205) Schönberg, A., Mamluk, M.: Tetrahedron Lett. *1971*, 4993; revised in Tetrahedron Lett. *1972*, 1512 by the same authors.
206) Marmor, S., Thomas, M. M.: J. Org. Chem. *32*, 252 (1967).
207) Crandall, J. K., Conover, W. W.: Tetrahedron Lett. *1971*, 583.
208) Lown, J. W., Maloney, T. W.: J. Org. Chem. *35*, 1716 (1970).
209) Dehmlow, E. V.: Liebigs Ann. Chem. *729*, 64 (1969).
210) Perkins, W. C., Wadsworth, D. H.: J. Org. Chem. *37*, 800 (1972).
211) Perkins, W. C., Wadsworth, D. H.: Synthesis *1972*, 205.
212) Toda, F., Akagi, K.: Tetrahedron Lett. *1968*, 3735.
213) Agranat, I., Cohen, S.: Bull. Chem. Soc. Japan *47*, 723 (1974).
214) Ciabattoni, J.: private communication cited in Ref.[192].
215) Naruse, M., Utimoto, K., Nozaki, H.: Tetrahedron Lett. *1972*, 1605.
216) Dunkelblum, E.: Tetrahedron *28*, 3879 (1972).
217) Bird, C. W., Harmer, A. F.: Organic Preparations and Procedures International *2*, 79 (1970).
218) Agranat, I., Pick, M. R.: Tetrahedron Lett. *1972*, 3111.
219) Eicher, Th., Böhm, S.: unpublished results.
220) Toda, F., Kataoka, T., Akagi, K.: Bull. Chem. Soc. Japan *44*, 244 (1971).
221) Walter, W., Krohn, J.: Liebigs Ann. Chem. *752*, 136 (1971).
222) Toda, F., Mitote, T., Akagi, K.: J. C. S. Chem. Commun. *1969*, 228.
223) Toda, F., Mitote, T., Akagi, K.: Bull. Chem. Soc. Japan *42*, 1777 (1969).
224) Dehmlow, E. V.: Tetrahedron Lett. *1967*, 5177.
225) Dehmlow, E. V.: Chem. Ber. *102*, 3863 (1969).
226) Gilchrist, T. L., Harris, C. J., Rees, C. W.: J. C. S. Chem. Commun. *1974*, 487.
227) Ciabattoni, J., Kocienski, P. J., Melloni, G.: Tetrahedron Lett. *1969*, 1883.
228) In addition to products *337/338* an adduct arising from two moles of diphenyl cyclopropenone was obtained (Ref.[227]).
229) Obata, N., Takizawa, T.: Tetrahedron Lett. *1970*, 2231.
230) Phenyl methoxy cyclopropenone is reported [Chickos, J. S.: J. Org. Chem. *38*, 3642 (1973)] to react with the above isocyanide in the same manner, but without assistance of triphenyl phosphine.
231) Hamada, A., Takizawa, T.: Tetrahedron Lett. *1972*, 1849.
232) Eicher, Th., Harth, R.: unpublished results.

233) For a recent review on cyclopropanone chemistry see Wasserman, H. H., Clark, G. C., and Turley, P. C.: Topics Curr. Chem./Fortschr. Chem. Forsch. *47*, 73 (1974).

234) Lown, J. W., Maloney, T. W., Dallas, G.: Can. J. Chem. *48*, 584 (1970).

235) Grigg, R., Jackson, J. L.: J. Chem. Soc. C *1970*, 552.

236) Sauer, J., Krapf, H.: Tetrahedron Lett. *1969*, 4279.

237) Steinfels, M. A., Dreiding, A. S.: Helv. Chim. Acta *55*, 702 (1972).

238) Bilinski, V., Steinfels, M. A., Dreiding, A. S.: Helv. Chim. Acta *55*, 1075 (1972).

239) a) Bilinski, V., Dreiding, A. S.: Helv. Chim. Acta *55*, 1271 (1972).
    b) β-Carbonylenamines in principle react analogously to enamines: Bilinski, V., Dreiding, A. S., Helv. Chim. Acta *57*, 2525 (1974).

240) Steinfels, M. A., Krapf, H. W., Riedl, P., Sauer, J., Dreiding, A. S.: Helv. Chim. Acta *55*, 1759 (1972); Riedl, P.: Dissertation Regensburg 1974.

241) Eicher, Th., Böhm, S.: Tetrahedron Lett. *1972*, 2603.

242) Eicher, Th., Böhm, S.: Chem. Ber. *107*, 2186 (1974).

243) Eicher, Th., Böhm, S.: Chem. Ber. *107*, 2215 (1974).

244) Reinhoudt, D. N., Kouwenhoven, C. G.: Tetrahedron Lett. *1973*, 3751.

245) Eicher, Th., Böhm, S.: Tetrahedron Lett. *1972*, 3965.

246) Eicher, Th., Böhm, S.: Chem. Ber. *107*, 2238 (1974); see also Lown, J. W., Maloney, T. W.: Chem. and Ind. *1970*, 870.

247) Eicher, Th., Weber, J. L.: Tetrahedron Lett. *1974*, 3409.

248) Rosen, M. H., Fengler, J., Bonet, G.: Tetrahedron Lett. *1973*, 949.

249) Franck-Neumann, M.: Tetrahedron Lett. *1966*, 341.

250) Lown, J. W., Matsumoto, K.: Can. J. Chem. *49*, 1165 (1971).

251) Eicher, Th., Weber, J. L.: Tetrahedron Lett. *1974*, 1381.

252) Eicher, Th., Weber, J. L.: Tetrahedron Lett. *1973*, 1541.

253) Eicher, Th., Weber, J. L.: unpublished results.

254) Hassner, A., Kascheres, A.: J. Org. Chem. *37*, 2328 (1972).

255) Grigg, R., Hayes, R., Jackson, J. L., King, T. J.: J. C. S. Chem. Commun. *1973*, 349.

256) Sasaki, T., Kanematsu, K., Yukimoto, Y., Kato, E.: Synth. Commun. *3*, 249 (1973).

257) Eicher, Th., Angerer, v., E.: unpublished results.

258) Franck-Neumann, M., Buchecker, C.: Tetrahedron Lett. *1969*, 2659.

259) Potts, K. T., Baum, J. S.: J. C. S. Chem. Commun. *1973*, 833.

260) Eicher, Th., Schäfer, V.: Tetrahedron 1974, *30*, 4025 (1974).

261) Matsukubo, H., Kato, H.: J. C. S. Chem. Commun. *1974*, 412.

262) Eicher, Th., Schäfer, V.: unpublished results.

263) Lown, J. W., Matsumoto, K.: Can. J. Chem. *50*, 534 (1972).

264) Lown, J. W., Smalley, R. K., Dallas, G., Maloney, T. W.: Can. J. Chem. *48*, 89 (1970).

265) Lown, J. W., Matsumoto, K.: Can. J. Chem. *48*, 3399 (1970).

266) Lown, J. W., Smalley, R. K.: Tetrahedron Lett. *1969*, 169.

267) Lown, J. W., Westwood, R., Moser, J. P.: Can. J. Chem. *48*, 1682 (1970).

268) Eicher, Th., Hansen, A.: Tetrahedron Lett. *1967*, 1169.

269) Eicher, Th., Angerer, v., E., Hansen, A.: Liebigs Ann. Chem. *746*, 102 (1971).

270) Hayasi, Y., Nozaki, H.: Tetrahedron *27*, 3085 (1971).

271) Eicher, Th., Helgert, A.: unpublished results.

272) Eicher, Th., Angerer, v., E.: Liebigs Ann. Chem. *746*, 120 (1971).

273) Sasaki, T., Kanematsu, K., Kakehi, A., Ito, G.: Tetrahedron *28*, 4947 (1972).

274) Sasaki, T., Kanematsu, K., Kakehi, A.: J. Org. Chem. *36*, 2451 (1971).

275) Lown, J. W., Matsumoto, K.: Can. J. Chem. *50*, 584 (1972).

276) Potts, K. T., Elliott, A. J., Sorm, M.: J. Org. Chem. *37*, 3838 (1972).

277) Bird, C. W., Hudec, J.: Chem. and Ind. *1959*, 570.

278) Bird, C. W., Hollins, E. M.: Chem. and Ind. *1964*, 1362.

279) Ayrey, G., Bird, C. W., Briggs, E. M., Harmer, A. F.: Organomet. Chem. Synth. *1*, 187 (1971).

280) Bird, C. W., Briggs, E. M., Hudec, J.: J. Chem. Soc. C *1967*, 1862.

281) Bird, C. W., Briggs, E. M.: J. Chem. Soc. A *1967*, 1004.

282) Fichteman, W. L., Schmidt, P., Orchin, M.: J. Organomet. Chem. *12*, 249 (1968).

283) Visser, J. P., Ramakers-Blom, J. E.: J. Organomet. Chem. *44*, C 63 (1972).

284) An analogous complex was obtained from methyl thiirene dioxide: Visser, J. P., Leliveld, C. G., Reinhoudt, D. N.: J. C. S. Chem. Commun. *1972*, 178.

285) Wong, W., Singer, S. J., Pitts, W. D., Watkins, S. F., Baddley, W. H.: J. C. S. Chem. Commun. *1972*, 672.

286) Prinzbach, H., Fischer, U.: Angew. Chem. *78*, 642 (1966); Angew. Chem. Int. Ed. Engl. *5*, 602 (1966).

287) Prinzbach, H., Fischer, U.: Chimia *20*, 156 (1966); Prinzbach, H., Fischer, U.: Helv. Chim. Acta *50*, 1669 (1967).

288) Murata, I., Takatsuki, K., Hara, O.: Tetrahedron Lett. *1970*, 4665.

289) Andreades, S.: C. A. *70*, P 19827 s (1969).

290) Eicher, Th., Berneth, H.: unpublished results.

291) Niizuma, S., Konishi, S., Kokubun, H., Koizumi, M.: Chem. Lett. *1972*, 643; Bull. Chem. Soc. Japan *46*, 2279 (1973).

292) Hecht, S. S: Tetrahedron Lett. *1972*, 3731.

293) Prinzbach, H., Fischer, U.: Helv. Chim. Acta *50*, 1692 (1967).

294) Gompper, R., Seybold, G.: The Jerusalem Symposium on Quantum Chemistry and Biochemistry, Vol. III, p. 215. Jerusalem: The Israel Academy of Sciences and Humanities 1971.

295) Ciabattoni, J., Nathan, III, E. C.: J. Amer. Chem. Soc. *89*, 3081 (1967).

296) Eicher, Th., Born, Th.: Tetrahedron Lett. *1970*, 981.

297) Eicher, Th., Born, Th.: Tetrahedron Lett. *1970*, 985.

298) Eicher, Th., Born, Th.: Liebigs Ann. Chem. *762*, 127 (1972).

299) Eicher, Th., Pfister, Th.: Tetrahedron Lett. *1972*, 3969.

300) Eicher, Th., Pfister, Th.: unpublished results.

301) Battiste, M. A., Sprouse, C. A.: Chem. and Eng. News *47* (40), 52 (1969).

302) Eicher, Th., Angerer, v., E.: Chem. Ber. *103*, 339 (1970).

303) Pyron, R. S., Jones, W. M.: J. Org. Chem. *32*, 4048 (1967).

Received September 25, 1974

The Higher Annulenones

Dr. Melvyn V. Sargent* and Dr. Terrence M. Cresp**

Department of Organic Chemistry, University of Western Australia, Nedlands, W. A. 6009, Australia

Contents

---

* Author to whom correspondence should be addressed.
** Present address: Chemistry Department, University College, Gordon Street, London, WC1H OAJ, England.

Introduction

The pioneering synthetic work of Sondheimer and his group has resulted in an understanding of the properties of completely conjugated monocarbocyclic polyenes (annulenes) and polyenynes (dehydroannulenes)[1]. This field has been extended to incorporate bridged annulenes notably by the contributions of Boekelheide[2] and Vogel[3].

In agreement with the Hückel rule those annulenes and dehydroannulenes which contain $(4 n + 2) \pi$ electrons and a reasonably planar carbon skeleton appear to be aromatic. Aromaticity in annulenes is usually equated with positive resonance energy and the absence of bond alternation. The most direct method of measuring bond alternation is by single crystal X-ray diffraction. Unfortunately this method has been applied in only a few cases.

The diamagnetic susceptibilities of cyclic molecules with delocalised π electron systems are unusually high[4], and this has been ascribed to the circulation of the π-electrons in an applied magnetic field. This concept of a ring current, although open to criticism on theoretical grounds[5], explains the experimental results[6]. The determination of exaltations of diamagnetic susceptibilities has been applied as a criterion of aromaticity[4]. An important consequence of the circulation of π-electrons in an applied magnetic field is that in the NMR spectrum the protons external to the ring are deshielded by the induced magnetic field and the internal protons are shielded. This effect is known as diatropicity[7], and its presence has often been equated with aromaticity. The assessment of a diamagnetic ring current from chemical shift data relies on the choice of a suitable model system which possesses as many as possible of the characteristics of the molecule in question except for the ring current. For example 1,3-cyclohexadiene is usually taken as a model for benzene[8].

Theoretical calculations indicate that the Hückel $(4 n + 2)$ rule should break down at higher values of n, with the onset of bond alternation[9], and it has been predicted that the limit should lie between 22- and 26-membered ring compounds[10]. However doubt has been cast on this prediction by the synthesis of a diatropic monodehydro[26]annulene[11] and a diatropic bisdehydro[30]annulene[12].

Using the SCF-MO method, Dewar and Gleicher[10], have calculated, assuming delocalised π bond systems for the lowest triplet states of the $4 n \pi$ electron annulenes, that they would all have negative resonance energies, except [12] and [16]-annulene which would have insignificantly positive resonance energies. This destabilisation by π electron delocalisation is referred to as antiaromaticity[13]. A consequence of the destabilisation of the $4 n$ annulenes by π electron delocalisation is bond alternation, and as n increases Dewar[14] has predicted that they should, like the $(4 n + 2)$ annulenes, converge to a non-aromatic limit.

The Hückel approximation predicts that the highest occupied orbitals of the $4 n$ π electron annulenes to be doubly degenerated and to be occupied by two electrons. The ground state of these annulenes might thus be a triplet with one electron in each orbital. These molecules are not paramagnetic so that the degeneracy has been removed probably by bond alternation. The levels are thus split and it is the presence of this very low-lying empty orbital which gives rise to a large paramagnetic contribution to the ring current[15-17]. The consequence of this is that the $4n$ annulenes will exhibit NMR spectra in which the internal protons will be at low field and the external protons will be at high field. This phenomenon is known as paratropicity[7]. The ring size limit for paratropicity in $4n$ annulenes has not yet been discovered.

"Annulenones" are fully conjugated monocarbocyclic ketones, and consequently they contain an odd number of carbon atoms. The nomenclature for annulenones[18] is an extension of that adopted for annulenes. Thus the number of carbon atoms in the ring is indicated by a prefix, [n]. Cycloheptatrienone (tropone) 1 is therefore [7]annulenone. Assuming the carbonyl group is polarised in the usual way it would be expected, by analogy with the annulenes, that those annulenones with a $(4n + 3)$-membered ring or $(4n + 2)\pi$ electrons would be diatropic, and those with a $(4n + 1)$-membered ring or $4n\pi$ electrons would be paratropic. Little theoretical work has been done on the higher annulenones. Using the simple HMO method Hess and co-workers[19] calculated that only the first two members of the annulenone series would have resonance energies which differ appreciably from zero. However the unreliability of the simple HMO theory when applied to systems containing heteroatoms or odd-membered rings has been demonstrated[14]. The high dipole moments (ca. 5 D) and basicities of substituted cyclopropenones[20, 21], and the low field resonance (τ 1.0) of the protons of cyclopropenone 3[22] [the first member of the $(4n + 3)$-annulenones] are usually explained by a large ground state contribution from the cyclopropenium oxide form 4 in which the positive charge is delocalised over the ring carbon atoms[23]. Tobey[24] has recently disputed this explanation.

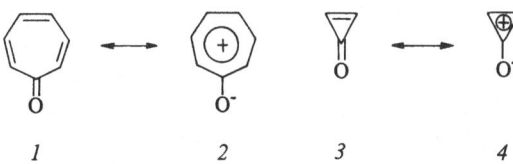

1   *2*   *3*   *4*

Cyclopentadienone 5, the first member of the $(4n + 1)$ annulenone series, is a highly unstable compound[25] undergoing dimerisation below $-80°$. The stable alkyl substituted compound 2,4-di-tert-butylcyclopentadienone 6 has the α-proton at τ 5.07 and the β-proton at 3.50 in its NMR spectrum[26]. These high field resonances could be due to an induced paramagnetic ring current but Garbisch and Sprecher[26] prefer to regard them as being due to an increase in π electron densities at the α- and β-positions of cyclopentadienone 5 relative to those of cyclopentenone since the chemical shifts of the tert-butyl protons of 6 are little different from those of 7 and 8.

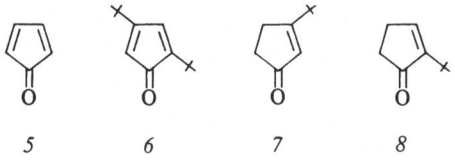

5   *6*   *7*   *8*

Tropone *1* [27] until recently the largest annulenone known, was regarded as being an aromatic molecule with a large ground state contribution from the dipolar form *2*. This appeared to be supported by its planarity[28], diamagnetic suceptibility[29], and its high dipole moment[30]. Bertelli[31] on the basis of dipole moment and NMR studies and CNDO/2 calculations suggested that tropone should behave as a planar cyclic polyenone with bond alternation. Calculations by Dewar[32] also support this view which is in accord with the recent X-ray crystal structure determination at −60°, which showed tropone to be a nearly planar molecule with distinct single and double bonds[33].

At present none of the parent higher annulenones ($n > 1$) have been reported. All the known annulenones are either polyenynones (dehydroannulenones), often with fused cyclohexene rings, or have pairs of internal hydrogens replaced by monatomic bridges. They will be discussed in order of increasing ring size.

$4n + 3, n = 2$

Most of our knowledge of [11]annulenones, which are potentially diatropic, comes from the researches of Vogel and his co-workers[34, 35]. By selenium dioxide oxidation of the cycloundecapentaene *9* or the 1,6-methano[11]annulenium ion *10* a mixture of the 4,9-, 3,8-, and 2,7-methano[11]annulenones *11–13* was obtained, and these were separated by chromatography. An alternative synthesis of *11* and *13* was by the acid-catalysed disproportionation of di-bicyclo[5.4.1]dodecapentaenyl ethers, which were generated from the ion *10* by treatment with sodium hydroxide.

The Vogel group has also developed rational synthesis of the above annulenones as well as of the other two possible isomers 2,8- *14* and 3,9-methano[11]annulenone *15*.

The rational synthesis of *11* started from 4,5-benzocycloheptenone ethylene ketal *16* which was reduced to the dihydrocompound *17* with lithium in liquid ammonia. Cyclopropanation of the latter with dichlorocarbene then gave the adduct *18*, the ketal oxygens of *17* presumably coordinating with the carbene and directing it

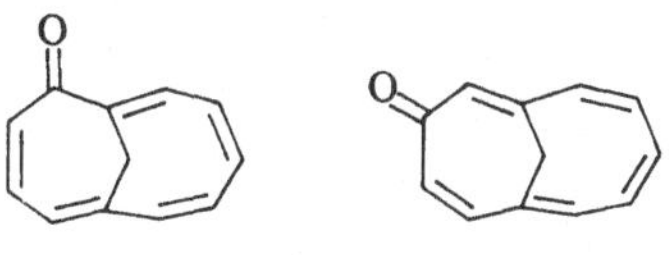

14   15

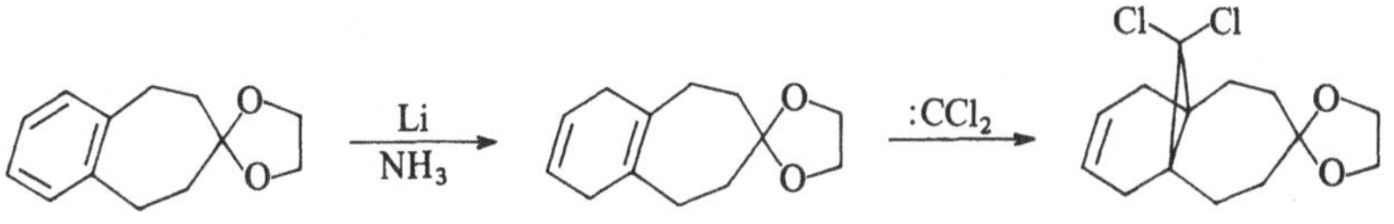

16   17   18

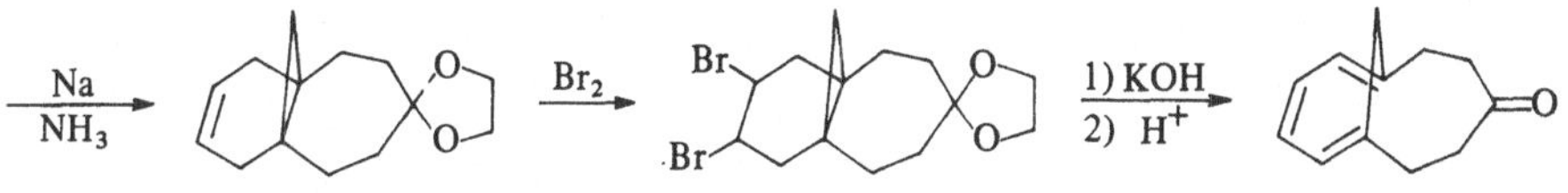

19   20   21

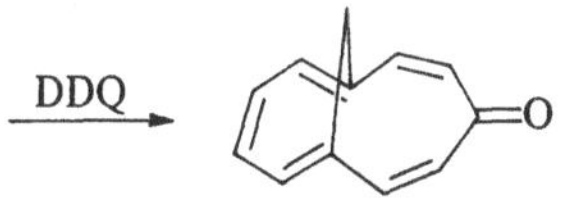

11

to the central double bond. Dechlorination of *18* was achieved with sodium in liquid ammonia and the resultant product *19* was treated with bromine at −70° and thus yielded the dibromo-compound *20*. Dehydrobromination of *20* occurred on heating it with potassium hydroxide in methanol and acidic treatment then gave the trienone *21*. This was dehydrogenated with 2,3-dichloro-5,6-dicyano-*p*-benzoquinone in benzene at 120° and yielded 4,9-methano[11]annulenone *11* as a stable yellow crystalline compound. An alternative approach to *11* involved condensation of cycloheptatriene-1,6-dicarbaldehyde *22* with acetone dicarboxylic ester *23* followed by hydrolysis and decarboxylation of the product *24*.

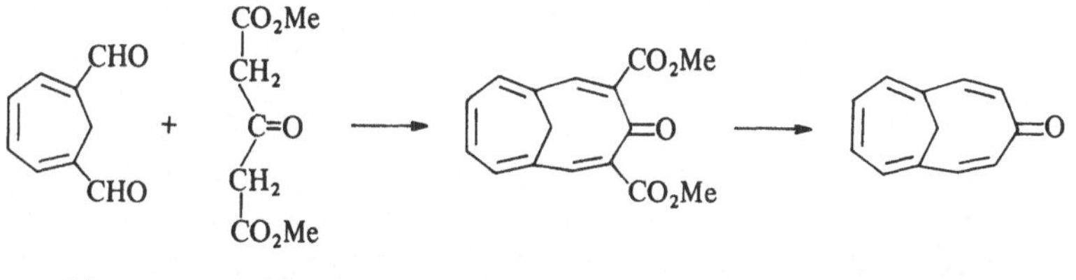

22   23   24   11

The rational synthesis of 3,8-methano[11]annulenone *12* started from 2-methoxy-1,4,5,8-tetrahydronaphthalene *25* which on reaction with dibromocarbene provided the bis-adduct *26*.

115

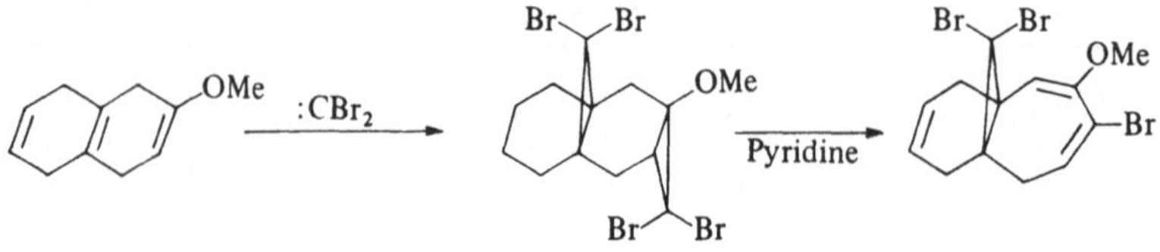

This on boiling in pyridine underwent opening of the methoxysubstituted cyclopropane ring exclusively and yielded *27*. Treatment of *27* successively with sodium in liquid ammonia and acid gave *28*. The remaining steps were analogous to those used in the synthesis of *11* and their application gave *12* as a moderately stable orange crystalline compound. A better route to *12* is the hydrogen iodide dechlorination of 11-chloro-3,8-methano[11]annulenone *31*.

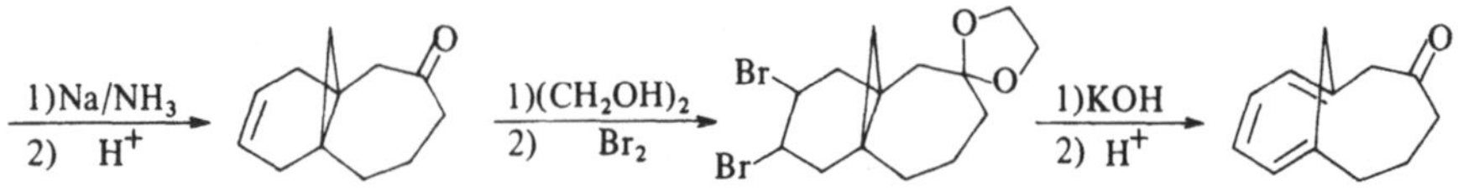

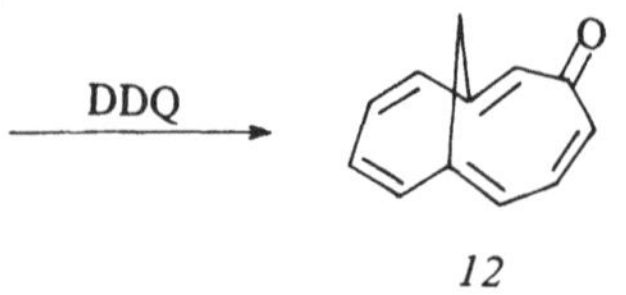

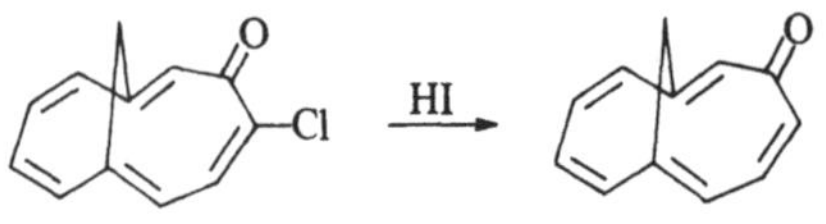

When 2-methoxy-1,6-methano[10]annulene *32* was subjected to a cyclopropanation with diazomethane and cuprous chloride as catalyst reaction occurred preferentially at the 5,6-, 6,7- and/or 1,10-bonds and the adducts spontaneously underwent disrotatory opening yielding the corresponding methoxybicyclo[5.4.1]dodecapentaenes. Hydride abstraction with triphenylmethyl fluoroborate was performed on the mixture and the ions *33* and *34* so produced were treated with dilute aqueous potassium hydroxide. The annulenones *13* and *14* were then separated by chromatography.

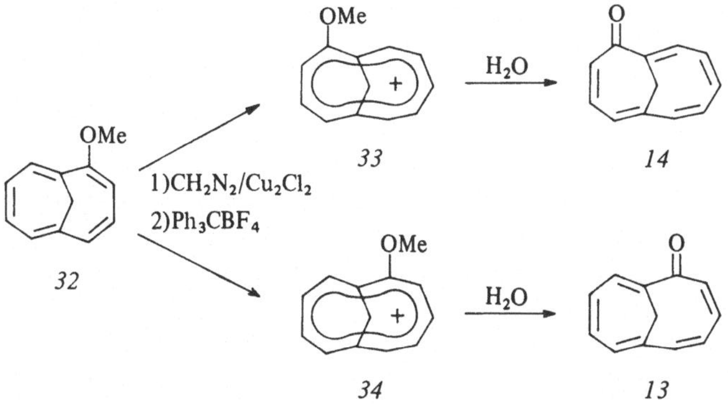

The same method was also applied to 3-tert-butoxy-1,6-methano[10]annulene which gave 3,8- *12* and 3,9-methano[11]annulenones *15*.

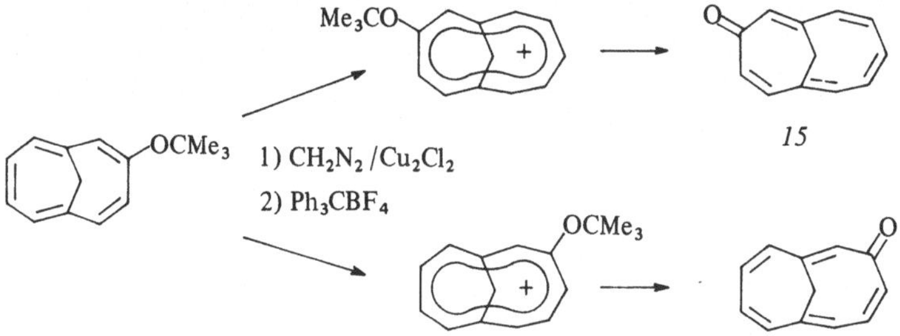

The NMR spectrum (CCl$_4$) of 4,9-methano[11]annulenone *11* exhibited an AA'XX' system at τ 2.99 and 3.10 (H$_6$, H$_7$ and H$_5$, H$_8$), an AB system at 2.82 and 3.98 with J 12.2 Hz (H$_3$, H$_{10}$ and H$_2$, H$_{11}$) and another AB system at 8.32 and 9.96 with J 11.4 Hz due to the *anti* H$_{12}$ (*anti* with respect to the cycloheptatriene ring) and *syn* H$_{12}$. A detailed analysis of the NMR spectrum gave the coupling constants: $J_{5,6} = J_{7,8}$ 6.74 Hz and $J_{6,7}$ 10.5 Hz. These coupling constants are very similar to those of the parent hydrocarbon bicyclo[5.4.1]dodeca-2,5,7,9,11-pentaene *36*, and this indicates that the ground state of the molecule is polyenone in nature. This conclusion has been verified by an X-ray structural determination on the annulenone which indicated alternation of bond lengths[36].

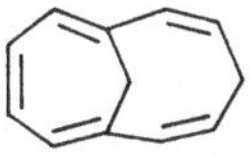

36

On deuteration the NMR spectrum (CF$_3$CO$_2$D) of *11* is markedly altered. The olefinic protons are shifted downfield and appear as an AB system with J 11.0 Hz at τ 1.03 and 1.98 (H$_3$, H$_{10}$ and H$_2$, H$_{11}$) and a singlet at 1.65 (H$_5$–H$_8$). The bridge protons are shifted upfield and appear as an AB system at τ 10.2 and 10.6 with J 11.0 Hz. These shifts are explained by the presence of an induced diamagnetic ring current in the ion *37* which causes the external protons to resonate at low field and the protons above the ring to resonate at high field.

37   11   38

The NMR spectra of the isomeric [11]annulenones *12–15* are more complex than that of *11* due to the lack of symmetry of these molecules. The same conclusions concerning the atropicity of the annulenones and the diatropicity of the hydroxyannulenium ions can, however, be drawn.

The basicities of the annulenones bear out the aromaticity of the hydroxyannulenium ions. The pK_a values for the five annulenones *11–15* fall in the range −0.7 to +0.6 (cf tropone −0.6) whereas ordinary ketones are in the range −6 to −7.

All the bridged [11]annulenones condense with malondinitrile in acetic anhydride and yield the corresponding dicyanomethanohendecafulvenes (*e.g. 11 → 38*) and these on treatment with strong acid furnish the substituted 1,6-methano[11]annulenium ions. 11-Chloro-3,8-methano[11]annulenone *31*, in parallel with the behaviour of the similarly 2-substituted tropone, undergoes reaction with sodium methoxide at room temperature to afford chiefly the products of normal *39* and ciné substitution *40*.

31   39   40

Electrocyclic reactions have been performed with three of the bridged [11]annulenones. Both *11* and *13*, which both contain a cycloheptatriene unit, undergo Diels-Alder additions with dienophiles *via* their norcaradiene valence tautomers *41* and *43* and yield adducts of the type *42* and *44*. Annulenone *13* was found to react only with 4-phenyl-1,2,4-triazoline-3,5-dione whilst *11* underwent reaction with a variety of dienophiles. 3,8-Methano[11]annulenone *12* contains a tetraene system and undergoes addition reactions, apparently of the (8 + 2)-type, at the termini of the tetraene system. Thus with maleic anhydride the adduct *46*, the valence tautomer of the initial adduct *45*, was isolated.

The tropolone vinylogue 2-hydroxy-4,9-methano[11]annulenone *52* has been synthesised. Tricyclo[4.4.1.0^{1,6}]undeca-3,8-diene *47* on cyclopropanation with dichlorocarbene afforded the monoadduct *48*, which on bromination-dehydrobromination gave the cycloheptatriene *49*. Under suitable conditions silver ion catalysed solvolysis of *49* gave the chloro-alcohol *50* in high yield as a single isomer, which underwent both oxidation and dehydrogenation with activated manganese dioxide and afforded *31*. This on heating in formic acid in presence of sodium formate at 140–150° gave the hydroxymethano[11]annulenone, either 11-hydroxy-3,8-methano[11]annulenone *51* or its tautomer 2-hydroxy-4,9-methano[11]annulenone *52* as a stable orange brown crystalline compound. Spectroscopic studies indicated

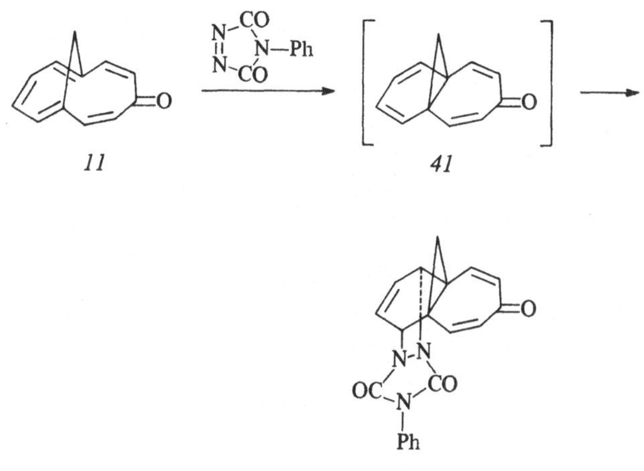

that the compound existed predominantly or entirely as the tautomer *52* which contains a cycloheptatriene structural unit. The hydroxyannulenone *52* exhibits amphoteric behaviour. It is almost as acidic as tropolone (pK$_a$ 6.9) having

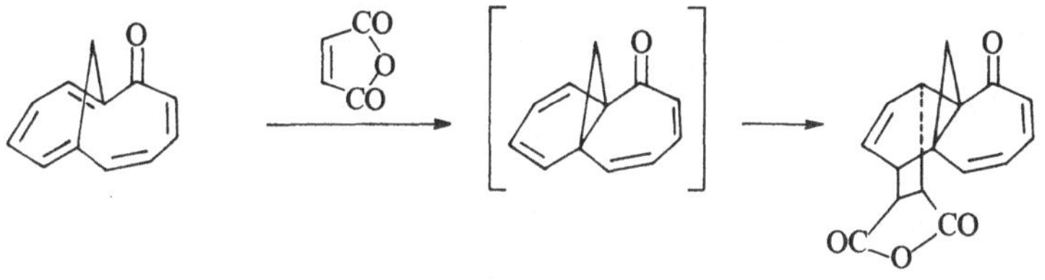

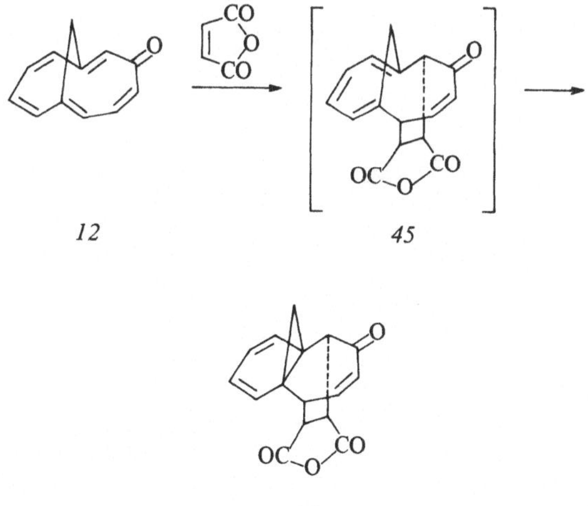

47 *48*

49 *50* *31*

51 *52*

53 *52* *54*

55 *56*

a pK_a of 8.2 and it reacts with alkali to give a salt *53*. With acid it gives the annulenium ion *54*. With diazomethane it gives the methyl ethers *55* and *56* in a ratio of 6 : 1.

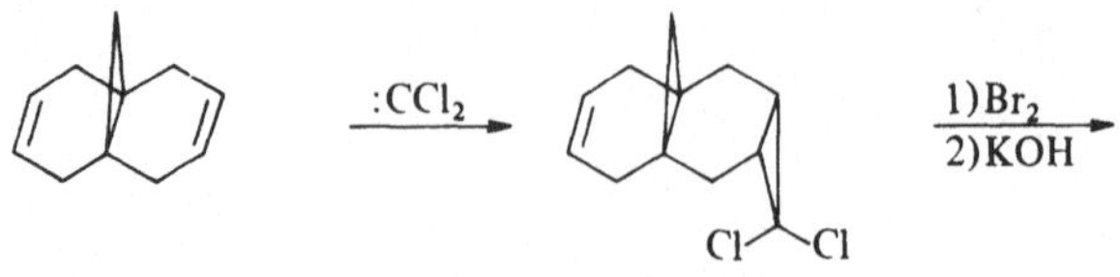

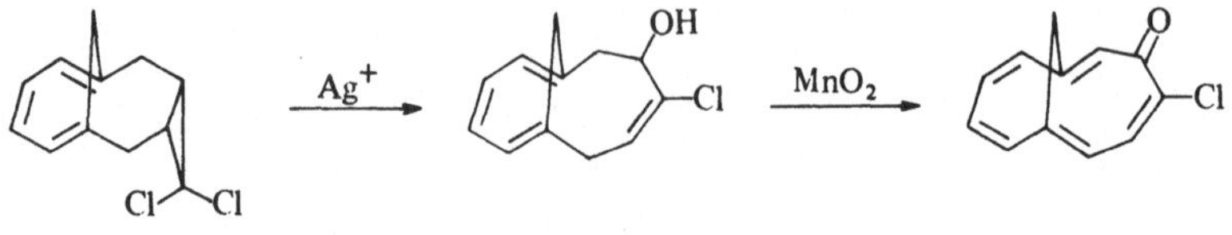

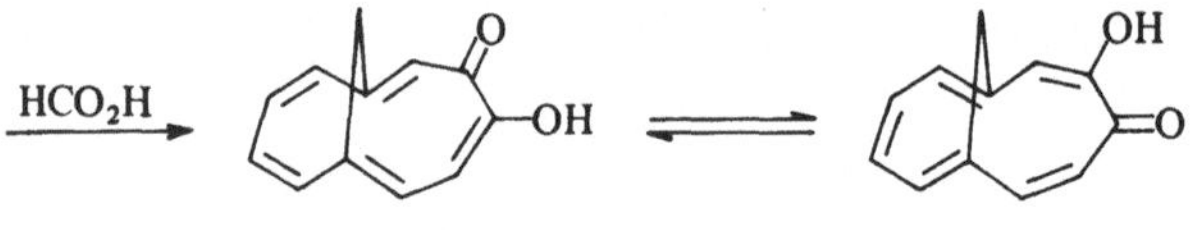

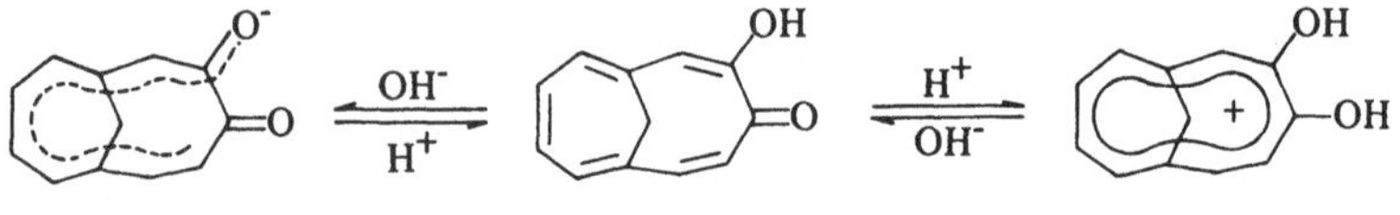

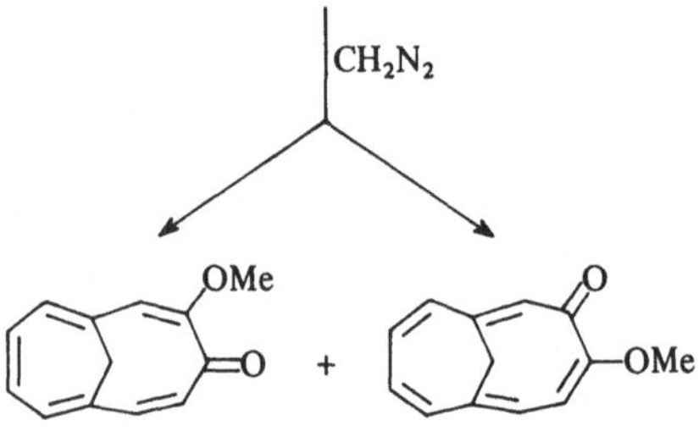

Ogawa and co-workers[37)] have reported the synthesis of 6,7-benzo-4,9-epoxy[11]-annulenone *60*. Manganese dioxide oxidation of the bis(hydroxymethylene)benzoxepin *57* gave the dialdehyde *58* which on condensation with dimethyl acetonedicarboxylate *23* in chloroform in presence of piperidinium acetate gave the diester *59*. This on hydrolysis followed by decarboxylation gave the annulenone *60*. The NMR spec-

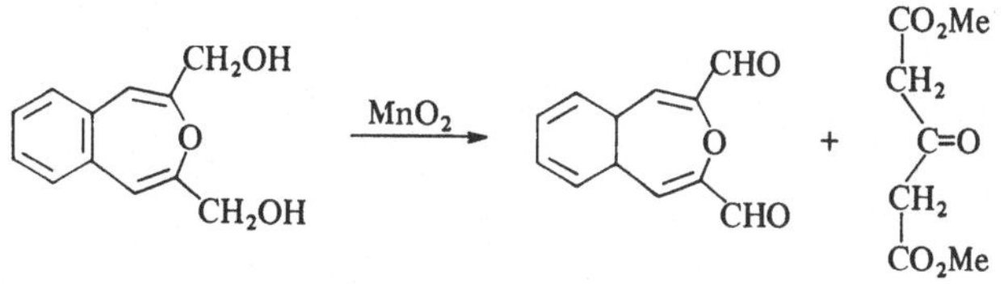

$$ 57 \quad\xrightarrow{MnO_2}\quad 58 \quad + \quad 23 $$

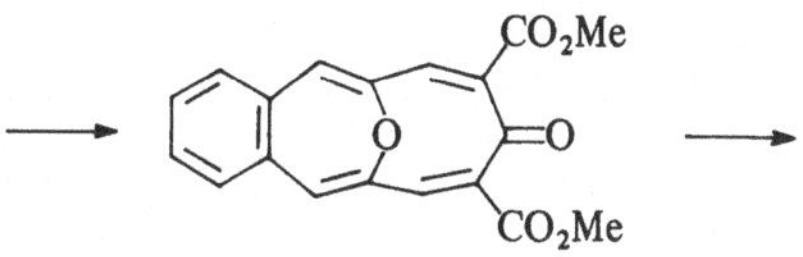

59

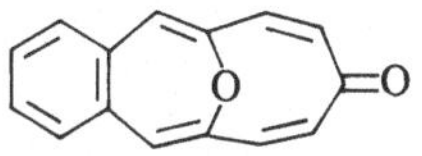

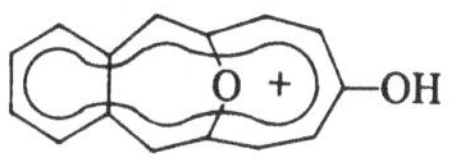

60   61

trum ($CDCl_3$) exhibited an AB system at τ 3.09 and 3.90 with J 11 Hz (H_3, H_{10} and H_2, H_{11}), a multiplet at 2.30 to 2.70 due to the benzenoid protons, and a singlet at 3.28 (H_5 and H_8). These data indicate that 60 is atropic. The electronic and NMR spectra of 60 in concentrated sulphuric acid have been interpreted in terms of the 14 π-electron ion 61.

$4n + 1, n = 3$

Four examples of this potentially paratropic group are known. By reaction of the mono-Grignard derivative of 1,2-diethynylcyclohexene 62 with ethyl formate Pilling and Sondheimer[38] obtained the alcohol 63 (27%) which under oxidative coupling with oxygen, cuprous chloride, and ammonium chloride in aqueous ethanol and benzene (Glaser conditions) yielded the cyclic alcohol 64 (40%). This on oxidation with manganese dioxide in ether at room temperature then gave the tetradehydro[13]-annulenone 65 as unstable red crystals, in high yield. Since 65 contains no protons bound directly to the 13-membered ring no firm conclusions can be drawn regarding the paratropicity of this molecule. It may be significant that the allylic protons of 65 resonate at τ 7.85 to 8.20 which is at higher field than the allylic protons of 64 or the alcohol 66, (obtained by treatment of the annulenone 65 with methylmagnesium iodide) which resonate at 7.65 to 8.00. The annulenone 65 formed a 2,4-dinitrophenylhydrazone and reacted with cyclopentadienyl anion to give the fulvalene

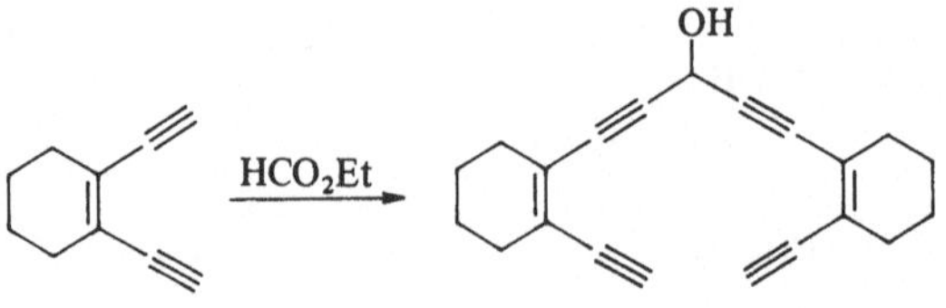

67. It however failed to react with methylenetriphenylphosphorane. Attempted partial hydrogenation of *65* using Lindlar's catalyst or 10% palladised charcoal led to none of the expected [13]annulenone *e.g. 68.* This may be due to the potential instability of the [13]annulenone which like its lower vinylogue cyclopentadienone *5* may undergo ready self-condensation.

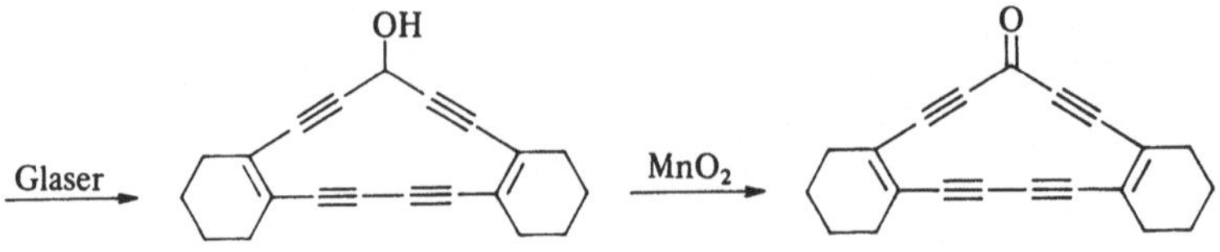

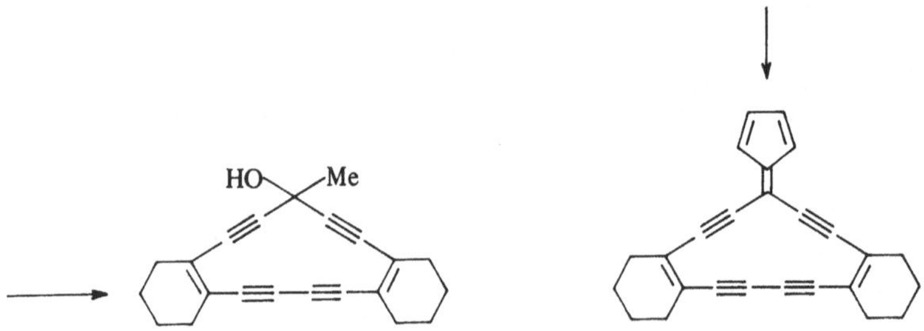

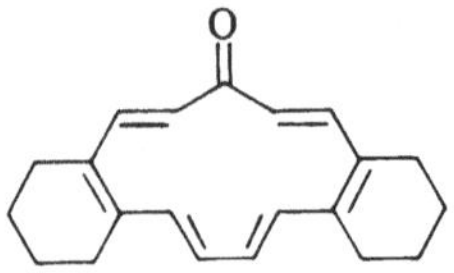

By condensation of 2-ethynylcyclohex-1-enylcarbaldehyde *69* with acetone under mild basic condition Howes, Le Goff, and Sondheimer[39] obtained the ketone *70.* The latter on oxidative coupling under Glaser or Eglington (cupric acetate monohydrate in pyridine) conditions gave the bisdehydro[13]annulenone *71* in 45 to 50% yield. The NMR assignments of both *70* and *71* were established by condensation of *69* with hexadeuterioacetone to yield the α,α'-dideuterio analogue of *70* which on oxidative coupling gave the similar analogue of *71*. In *70* the olefinic protons resonate as an AB system with J 16 Hz at τ 3.63 (H$_2$) and 2.02 (H$_3$) and in the annulenone *71* at

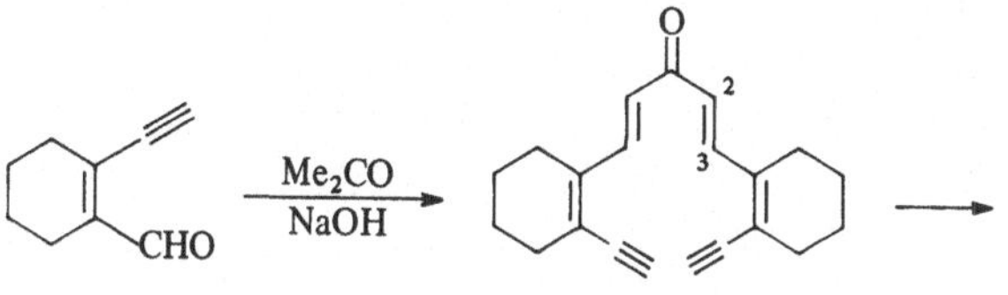

69   *70*

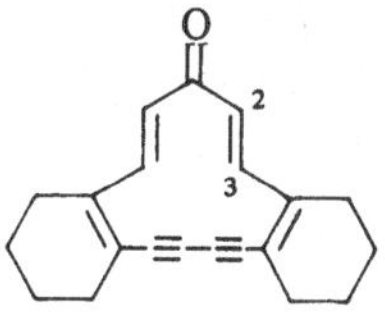

71

3.77 (H$_2$) and 0.77 (H$_3$). The downfield shift of H$_3$ on going from *70* to *71* is not due to a paramagnetic ring current in *71* as there is no significant upfield shift of H$_2$. Sondheimer and co-workers attribute the low field resonance of H$_3$ in *71* as being due to the anisotropy of the triple bond(s)[40]. Steric compression of the internal protons may also contribute to this downfield shift as observed in the [21]annulenone *139* (see later). It thus appears that the [13]annulenone *71* is atropic. The NMR spectrum of the deuteronated form of *71* obtained by dissolving *71* in deuterio-trifluoracetic acid exhibited an AB system with τ 3.51 (H$_2$) and −0.68 (H$_3$). Since both H$_2$ and H$_3$ have moved downfield compared with *71* this effect is due mainly to the positive charge and not due to paratropicity. The atropicity of the annulenone *71* is a reflection of its lack of planarity due to the strain or steric effects caused by the cyclohexane rings.

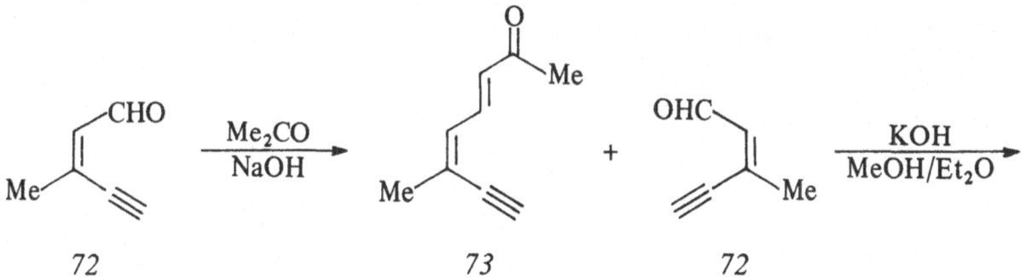

72   *73*   *72*

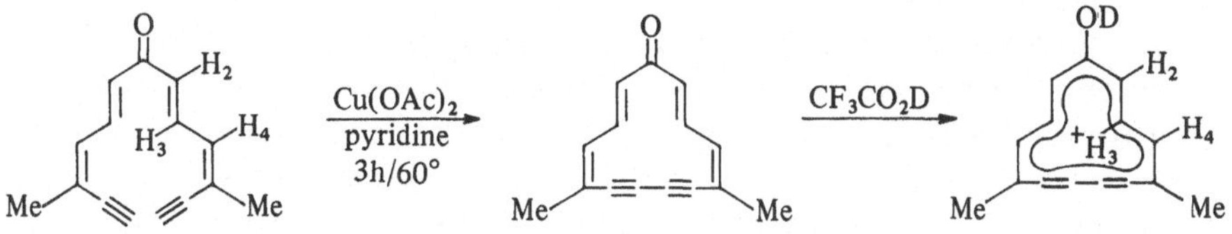

74   *75*   *76*

By contrast the recently synthesised dimethylbisdehydro[13]annulenone *75* appears to be paratropic[41]. Condensation of the aldehyde *72* with acetone under mild basic conditions gave the ketone *73* in 55% yield. This was subjected to a base catalysed condensation with *72* and yielded the ketone *74* in 41% yield. Eglington coupling of the latter gave the annulenone *75* in 28% yield. In the acyclic ketone *74* the olefinic protons occur as a doublet at τ 3.55 (H_2, J 16 Hz), a doublet of doublets at 2.32 (H_3, J 16, 11 Hz), and a doublet at 3.54 (H_4, J 11 Hz), and the methyl protons occur as a singlet at 7.98. In the annulenone *75*, however, the olefinic protons occur as a doublet at τ 3.90 (H_2, J 17 Hz), a doublet of doublets at 0.61 (H_3, J 17, 10 Hz) and a doublet at 3.71 (H_4, J 10 Hz), and the methyl protons as a singlet at 8.26. These data clearly indicate the paratropicity of *75*. In deuteriotrifluoroacetic acid the paratropic hydroxy[13]annulenium ion *76* is produced which exhibits in its NMR spectrum a doublet at τ 3.85 (H_2, J 16 Hz), a doublet of doublets at –0.79 (H_3, J 16, 10 Hz), a doublet at 3.88 (H_4, J 10 Hz), and a singlet due to the methyl protons at 8.33.

Nakagawa and co-workers[42] by an analogous series of reactions have synthesised the di-tert-butylbisdehydro[13]annulenone *77* which exhibits similar properties to the dimethyl analogue *75*.

77

4 n + 3, n = 3

Annulenones containing 15 carbon-atoms may be expected to be diatropic. Since the [13]annulenones were paratropic it was of interest to determine the NMR spectral properties of the next higher members.

Reaction of 1,2-diethynylcyclohexene *62* with 1 molar equivalent of ethyl-magnesium bromide followed by treatment with N,N-dimethylformamide in tetrahydrofuran gave the acetylenic aldehyde *78*[43]. This aldehyde on reaction with the mono-Grignard reagents of the unsymmetrical diacetylene *79* gave a mixture of the isomeric alcohols *80* and *81*. Oxidative coupling of the mixture under Glaser conditions afforded the extremely unstable cyclic alcohol *82* as the sole monomeric product which was immediately oxidised with manganese dioxide in ether and gave the tetradehydro[15]annulenone *83*. Treatment of *83* with methyl-lithium gave the tertiary alcohol *84*. In the NMR spectrum of *83* the external protons resonate at lower field [τ 3.55 (H_4), 2.35 (H_6), and 3.72 (H_7)] than the analogous protons of *84* [τ 4.23 (H_4), 3.17 (H_6), and 4.43 (H_7)] and the internal proton of *83* [τ 4.99 (H_5)] resonates at higher field than the internal proton of *84* [τ 3.03 (H_5)] thus demonstrating the diatropicity of *83*. On deuteronation of *83* the ring-current effect is enhanced and in the NMR spectrum of *83* (CF_3CO_2D) the external protons $(H_4, H_6,$ and $H_7)$ resonate at τ 1.93 to 3.26 and the internal proton (H_5) at 6.51.

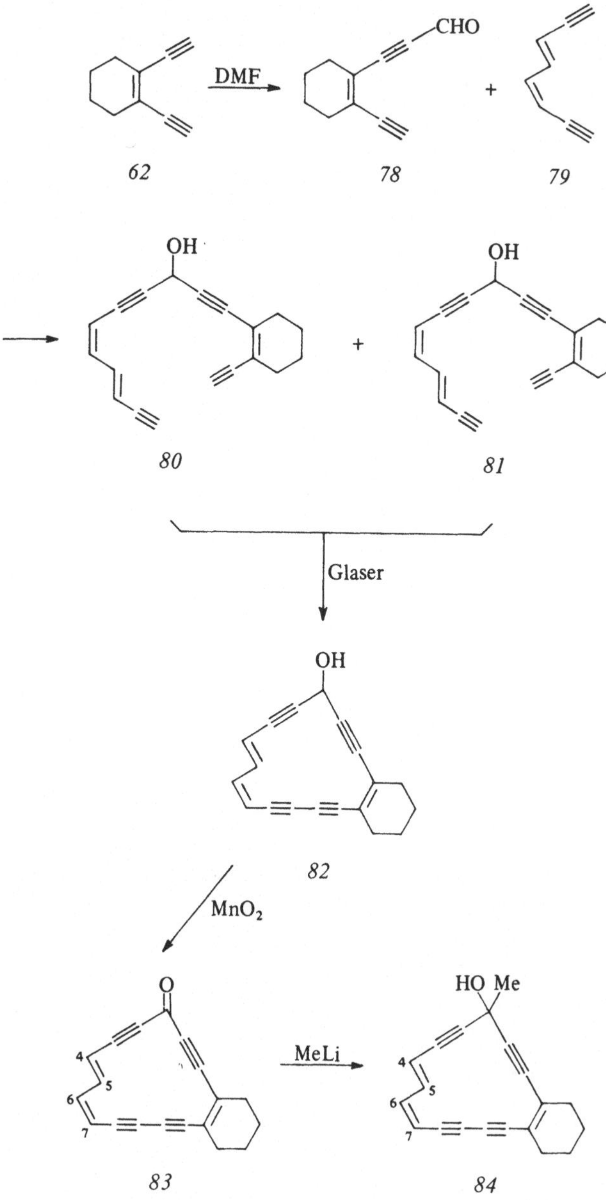

Condensation of 2-ethynylcyclohex-1-enylcarbaldehyde *69* with an excess of acetone gave the ketone *85*[44]. Reaction of equimolar amounts of *85* and *86* in ethereal methanolic potassium hydroxide gave the ketone *87*. Oxidative coupling of *87* under Glaser conditions afforded two separable isomeric bisdehydro[15]-annulenones *88* and *89*. The mono-*cis* isomer *88* may have the structure *90* in which

125

the double bond on the other side of the carboxyl group is *cis* but the former configuration is preferred on examination of molecular models. In the NMR spectra of

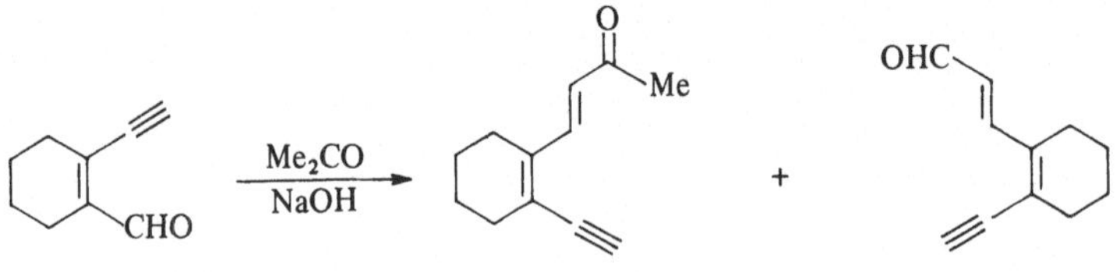

69   85   86

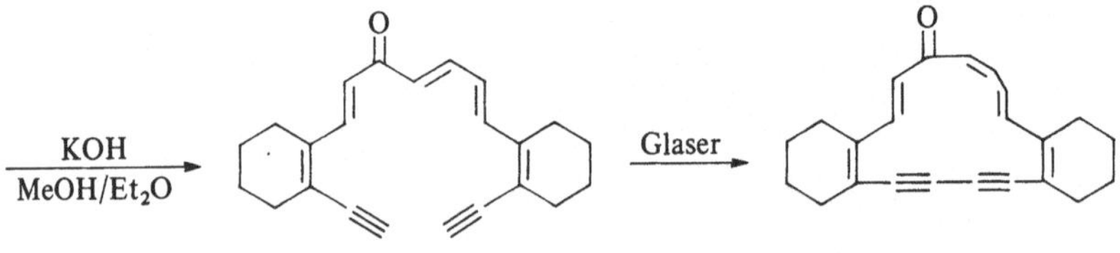

87   88

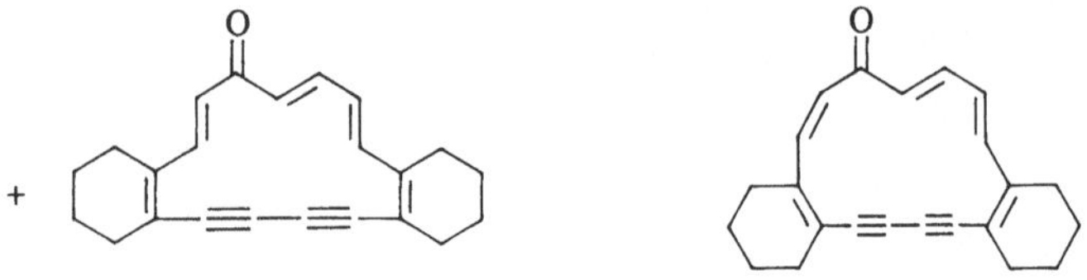

89   90

88 and *89* no clear distinction can be made between external and internal olefinic protons and comparison with the NMR spectrum of the acyclic diacetylene *87* indicated that both *88* and *89* were atropic. On deuteronation of either *88* or *89* with deuteriotrifluoroacetic acid the same deuterated species *91* is formed. In its NMR

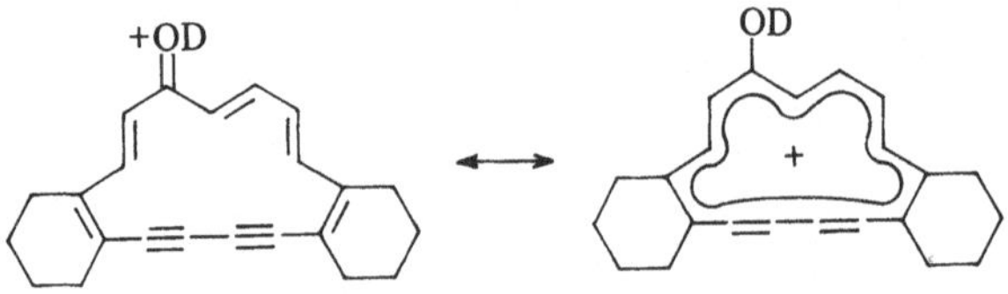

91

spectrum *91* exhibits resonances due to the internal protons at very high field (τ 9.6 to 9.9) and those due to the external protons at very low field (τ 0.4 to 1.5) demonstrating that the protonated species *91* is diatropic. On quenching with water *91* gives *89* irrespective of whether it was formed from *88* or *89*.

    By base catalysed condensation of the aldehyde *92* with the ketone *73* Sondheimer and Ojima[41] secured a 42% yield of the ketone *93* which underwent cyclisation under Eglinton conditions to afford a 16% yield of the dimethylbisdehydro[15]-annulenone *94*. By comparison of the NMR spectral data for *93* and *94* it is apparent that *94* is diatropic. The diatropicity of *94* is, as expected, increased on protonation.

92   73   93

94

Condensation of *cis*-α,β-di(5-formyl-2-furyl)ethylene *95* with acetone dicarboxylic ester *23* in the presence of piperidium acetate gave the diester *96*[45] which on concentrated sulphuric acid treatment yielded the anhydride *97*. Decarboxylation of the derived diacid *98* with copper chromite in quinoline[46] gave the all-*cis* [15]annulenone *99* and the mono-*trans* isomer *100*. The annulenones *96*, *98*, *99*, and *100* are atropic, however in the NMR spectrum of the anhydride *97* the protons resonate at considerably lower field than those of the diester *96* indicating that *97* is diatropic. Molecular models suggest that the anhydride substituent holds the molecule in a rigid planar conformation. The protonated forms of *99* and *100* are both diatropic, thus in the NMR spectrum of *100* (CF_3CO_2H) the external protons (τ −0.39 to 1.07) and the internal proton (τ 13.56) are at dramatically different field.

Recently the diatropic oxygen-bridged [15]annulenone *104* has been synthesised[47]. Wittig reaction of the dicarbaldyhyde *101* with the phosphonium salt *102* gave the dihydro[15]annulenone *103* in 15% yield. This was monobrominated with *N*-bromosuccinimide and the crude product was then dehydrobrominated and yielded the annulenone *104* as stable orange-red needles. The NMR spectrum of *104* exhibited an AB system at τ 1.95 and 2.82 (H_3, H_{14}, H_4, H_{13}, J 3.6 Hz), a multiplet at 2.92 (H_6, H_7, H_{10} H_{11}), and a multiplet at 3.92 (H_8, H_9). The diatropicity of *104* followed from a comparison with the NMR spectrum of *103* in which the furanoid protons resonate as an AB system at τ 2.73 and 3.78 (H_3, H_{14}, H_4, H_{13}, J 3.5 Hz). Annulenone *104* is a conformationally mobile system due to the rapid rotation of the *trans*-double bond. The NMR spectrum of *104* (CD_2Cl_2/CS_2) was found to be temperature dependent and on cooling only the high field multiplet lost its fine structure: at −80° it was a very broad signal $\left(\dfrac{Wh}{2} \ 30 \ \text{Hz}\right)$, and at −90° it could not longer be discerned.

127

95 *23* *96*

97 *98*

99 *100*

101 *103* *104*

+

$Br\ Ph_3\overset{+}{P}(CH_2)_4\overset{+}{P}Ph_3\ Br$

102

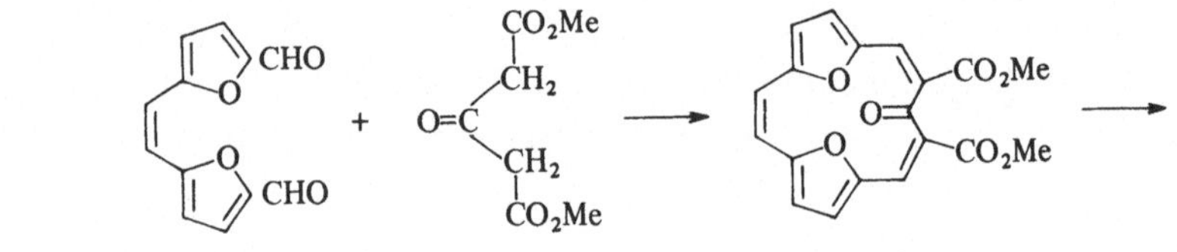

4*n* + 1, *n* = 4

Annulenones containing 17 carbon atoms and 16 π-electrons may be expected to be paratropic.

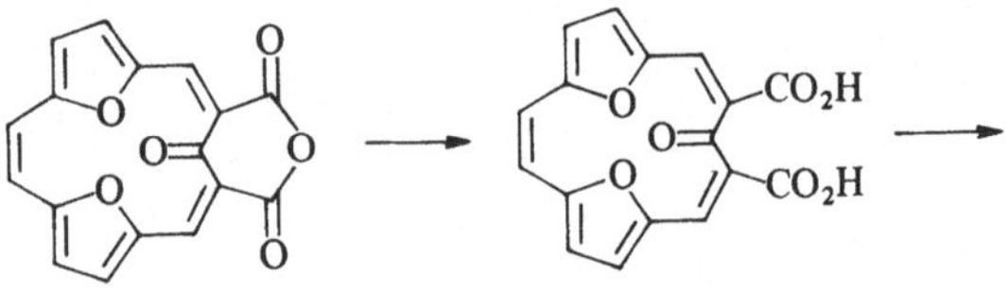

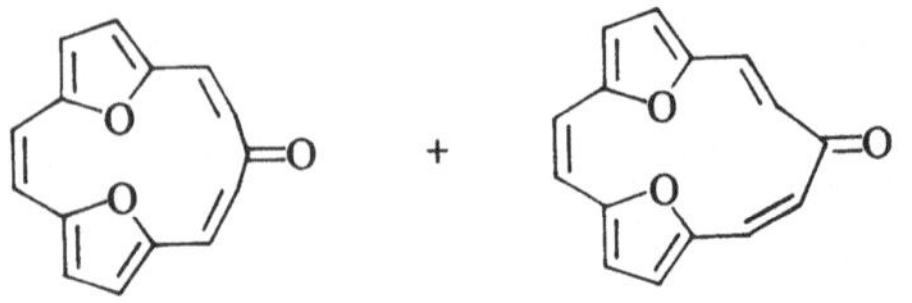

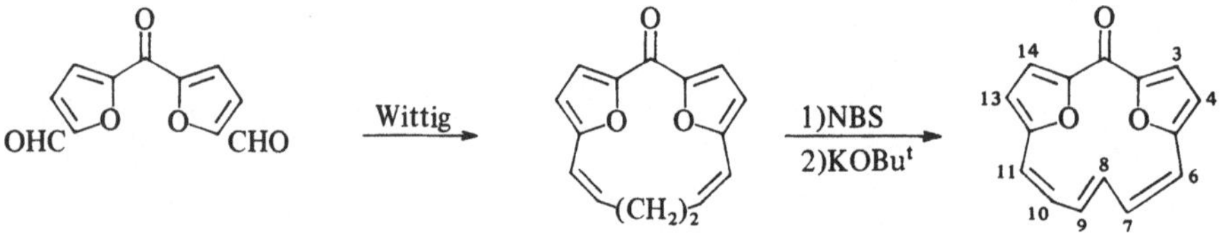

Treatment of the symmetrical diyne *105* with 1 molar equivalent of ethyl magnesium bromide followed by 0.5 molar equivalents of ethyl formate gave the di-*trans* alcohol *106* which on oxidative coupling under Glaser conditions gave the cyclic alcohol *107*[48]. The corresponding ketone *108* obtained by Jones oxidation of *107* was expected to yield a bisdehydro[17]annulenone *e.g.* *109* on prototropic rearrangement with potassium tert-butoxide in tetrahydrofuran, since similar rearrangements have been observed with cyclic 1,4-enynes[49]. However both rearrangement and dehydrogenation occurred. With undistilled tetrahydrofuran as solvent only the symmetrical tetradehydro[17]annulenone *110*[48] was formed whilst with freshly distilled tetrahydrofuran as solvent two new annulenones *110*[50] and *112*[1a] were isolated. That these [17]annulenones *110*, *111*, and *112* are paratropic was demon-

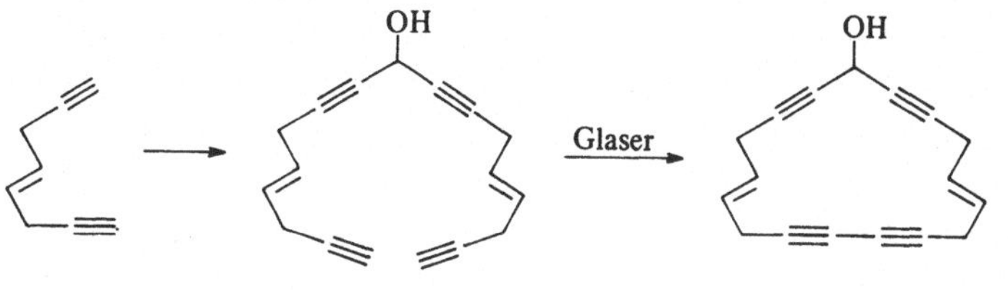

105 *106* Glaser *107*

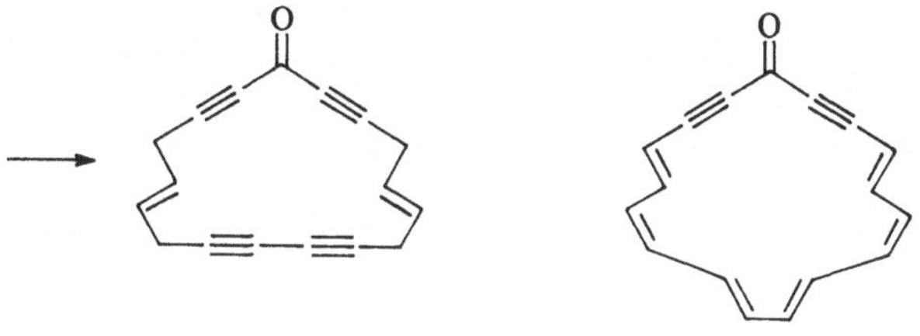

108 *109*

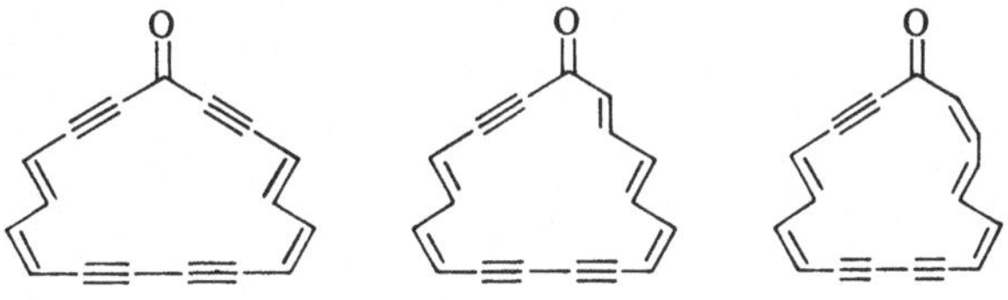

110 *111* *112*

strated by the low field resonances (τ ca. −0.5 to 0.4) of the internal protons and the high field resonances (τ ca. 3.9 to 5.0) of the external protons. It is well known that in annulenes it is the inner protons that are shifted further from the normal resonance position of olefinic protons than are the outer protons. Attempts to form

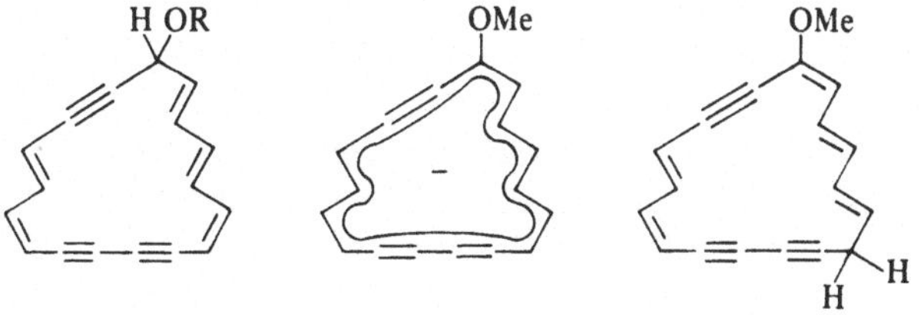

113 R = H
114 R = Me

115

116

[17]annulenone itself by catalytic partial hydrogenation of *111* and *112* were not successful[51]. Reduction of *112* with sodium borohydride[50] followed by methylation of the alcohol *113* gave the methyl ether *114*. In the NMR spectrum of *114* the internal protons resonate at τ 1.93 to 2.77 and the external protons resonate at 3.15 to 4.67 which is indicative of a small paramagnetic ring current in *114* reflecting the polarization of the carbon-oxygen bond. Treatment of *114* with methyl-lithium in perdeuteriotetrahydrofuran gave the anion *115* which had resonances in its NMR spectrum at τ 18.54 to 19.09 due to the internal protons and at −0.47 to 2.16 due to the external protons clearly demonstrating the diatropicity of this 18 π-electron system. Quenching of *115* with water yielded the isomeric ether *116* instead of *114*[50].

In an analogous manner to the synthesis of 4,5:10,11-bis(tetramethylene)-6,8-bisdehydro[13]annulenone *71* reaction of the α,β-unsaturated aldehyde *86* with acetone yielded the ketone *117* which on oxidative coupling under Glaser conditions gave the bisdehydro[17]annulenone *118*[39]. In contrast to the atropicity of the lower vinylogue *71* the [17]annulenone *118* is paratropic as in its NMR spectrum the external protons (H$_2$ and H$_4$) resonate at τ 4.14 and 3.92 respectively and the internal protons (H$_3$ and H$_5$) resonate at 1.42 and 1.15 respectively.

This difference is further enhanced when the NMR spectrum of *118* is recorded in deuteriotrifluoroacetic acid when the internal protons are shifted downfield to τ −2.65 and −2.85 and the external protons are shifted upfield to 4.38 and 4.47. In the NMR spectrum of the acylic precursor *117* the protons H$_2$, H$_3$, H$_4$, and H$_5$ resonate at τ 3.48, 2.45, 3.53, and 2.58 respectively.

By a similar synthetic sequence to that adopted for *118*, Sondheimer and Ojima[41] have synthesised the dimethyl analogue *119*. Again this proved to be paratropic, and the ring current is enhanced on deuteronation.

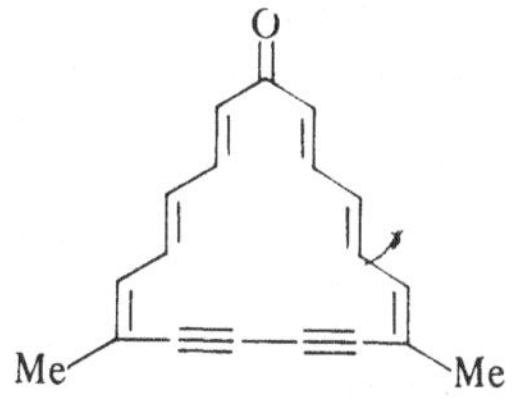

119

Wittig reactions of the bisphosphonium salt *120* with furan-2,5-dicarbaldehyde *121*, pyrrole-2,5-dicarbaldehyde *122* and thiophen-2,5-dicarbaldehyde *123* afforded the heteroatom-bridged [17]annulenones *124, 125* and *126* respectively[52]. These on reduction with lithium aluminium hydride-aluminium chloride gave the corresponding homoannulenes *127, 128,* and *129*. In their NMR spectra the external protons of *124* and *125* resonate at higher field than the analogous protons of the homoannulenes *127* and *128* respectively, indicative of a paramagnetic ring current in *124* and *125*. Furthermore the internal imino proton of *128* resonates at τ −2.4 and is shifted downfield to −8.3 in *125*. The low field resonance of the internal imino proton in the 16 π-electron system *125* is in marked contrast to that of the bridging imino proton of the 10 π-electron system 1,6-epimino[10]annulene *133*[53] which resonates at τ 11.1 and with the internal imino protons of the 18 π-electron porphyrin

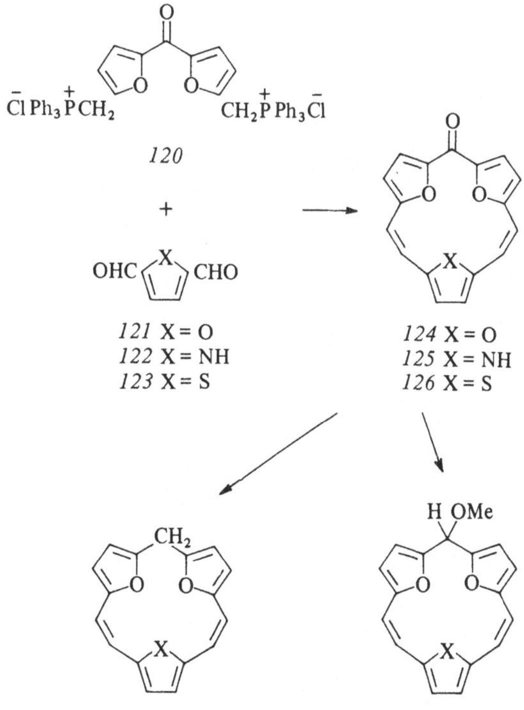

120

+

121 X = O
122 X = NH
123 X = S

124 X = O
125 X = NH
126 X = S

127 X = O
128 X = NH
129 X = S

130 X = O
131 X = NH
132 X = S

131

systems *134*[54)] which resonate at *ca.* τ 14. In the epithio[17]annulenone *126* the bulky sulphur atom presumably prevents the molecule from attaining planarity and

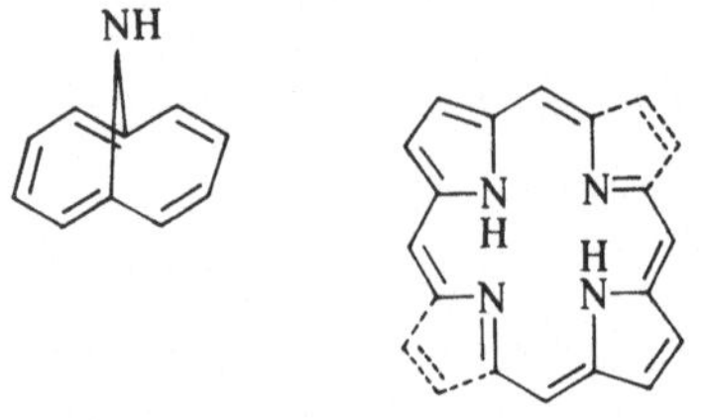

133   *134*

comparison of the proton chemical shifts of *126* with those of the model system *129* indicates that *126* fails to support any appreciable paramagnetic ring current. The presence of a thiophen unit in *126* may also introduce significant perturbations.

Reduction of *124, 125,* and *126* with either sodium borohydride or lithium aluminium hydride followed by methylation of the unstable alcohol obtained gave the methyl ethers *130, 131,* and *132* respectively. Comparisons between the NMR spectra of the methyl ethers *130* and *131* and the homoannulenes *127* and *128* indicated that *130* and *131* are paratropic like the 16 π-electron methyl ether *114*. The methyl ether *132* is atropic.

4 *n* + 3, *n* = 4

Only one example of the potentially diatropic [19]annulenones is known. Wittig reaction of 5-(β-formylvinyl)-2-furaldehyde *135* with the bisphosphonium salt *120*

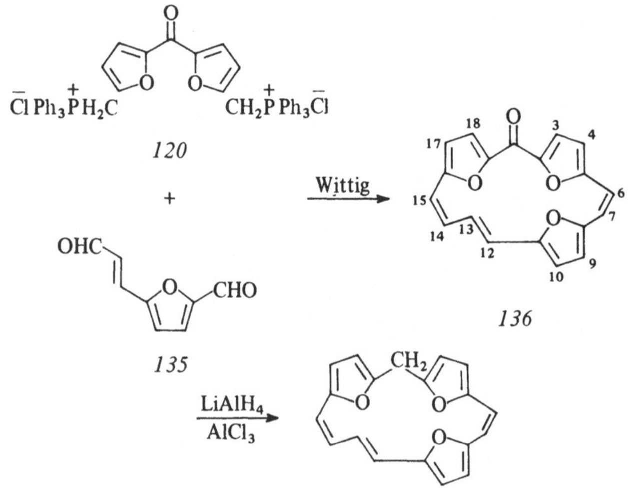

120

135

136

137

gave the diatropic [19]annulenone *136* as red prisms in low yield[55]. The NMR spectrum was highly complex and a partial first order analysis was made by the INDOR technique monitoring at the frequencies of H_3, H_{13}, and H_{18}. It exhibited an AB system with J 3.5 Hz at τ 1.86 and 2.87 (H_3, H_4 or H_{18}, H_{17}), an AB system at 2.07 and 2.95 with J 3.5 Hz (H_3, H_4 or H_{18}, H_{17}), a doublet with $J_{12, 13}$ 15.0 Hz, at 2.61 (H_{12}), a multiplet at 2.63 to 3.02 (H_6, H_7, H_9, H_{10}, and H_{15}), a doublet of doublets with $J_{13, 14}$ 11.5 Hz and $J_{14, 15}$ 11.2 Hz at 3.01 (H_{14}), and a doublet of doublets with $J_{13, 14}$ 11.5 Hz and $J_{12, 13}$ 15.0 Hz at 4.88 (H_{13}). The results indicate that the 12,13,14,15-diene system is *trans,cis* or *cis,trans*. The former was favoured on the assumption that the $\alpha\beta$-unsaturated aldehyde *135* retained its *trans*-stereochemistry on Wittig reaction. Should the stereochemistry be reversed the conclusion regarding the diatropicity of the annulenone *136* is unaltered. It was also assumed that the stereochemistry of the 6,7-double bond is *cis* since molecular models indicate that a cyclic molecule cannot be obtained if the 6,7-double bond is *trans*. Reduction of the annulenone *136* with lithium aluminium hydride/aluminium chloride gave the homoannulene *137* the NMR spectrum of which was not amenable to analysis. It exhibited a multiplet at τ 3.22 to 3.67 due to the furan and olefinic protons, and a singlet at 5.95 due to the methylene protons.

That the [19]annulenone *136* is diatropic is evident from the high field resonance of H_{13} compared to the adjacent external protons H_{12} and H_{14}, and to all the olefinic protons of the homoannulene *137*. Further evidence is the significantly lower field resonance of H_3, H_4, H_{17}, and H_{18} when compared with the similar protons H_3 and H_4 of the atropic ketones *143*, *144*, and *145* (see below).

$4n + 1, n = 5$

Three heteroatom bridged examples of this potentially paratropic 20 π-electron group are known[55]. Wittig reaction between the bisphosphonium salt *120* and 2,5-bis(β-formylvinyl)furan *138* gave a moderate yield of the symmetrical triepoxy[21]-annulenone *139*. On one occasion this Wittig reaction gave in addition a very low yield of another [21]annulenone *140* which was thought to arise by contamination of the aldehyde *138* with its isomer *142*. The NMR spectrum of annulenone *140* which was only poorly soluble in most organic solvents was highly complex and exhibited a 2 H multiplet at τ −1.2 to 0.8, and a 12 H multiplet at τ 2.7 to 4.7. The NMR spectral data are best accommodated by the unsymmetrical structure *140*. The very low field resonances of the internal protons may indicate that *140* is paratropic but in the absence of more concrete data no firm conclusions were drawn. Reduction of the annulenone *139* with lithium aluminium hydride-aluminium chloride gave the homoannulene *141*.

In the NMR spectrum of the annulenone *139* the internal protons (H_8) resonate at τ 1.95 and the external olefinic protons (H_6, H_7, and H_9) resonate at 3.7−4.0. That the low field resonance of the internal protons (H_8) is not due to the presence of a paramagnetic ring current in the annulenone *139* is demonstrated by the even lower field resonance (τ 0.89) of the analogous protons in the homoannulene *141*. This deshielding effect in *139* and *141* has been ascribed to the matual steric compression of the internal protons since in the model ketones *143, 144,* and *145* the internal protons (H_7) also resonate at appreciably lower field than the external olefinic protons (H_6, H_8, and H_9). As the ring size decreases the deshielding effect is enhanced, thus the internal protons (H_7) of ketone *145* resonate at τ 2.87, the internal protons of *144* resonate at 1.94, and the internal protons of 143 resonate at 1.49. The chemical shifts of the furan protons (H_3 and H_4) in annulenone *139* are very similar to those of the similar protons (H_3 and H_4) of the model ketones *143, 144,* and *145*. It was therefore concluded that the annulenone *139* is atropic.

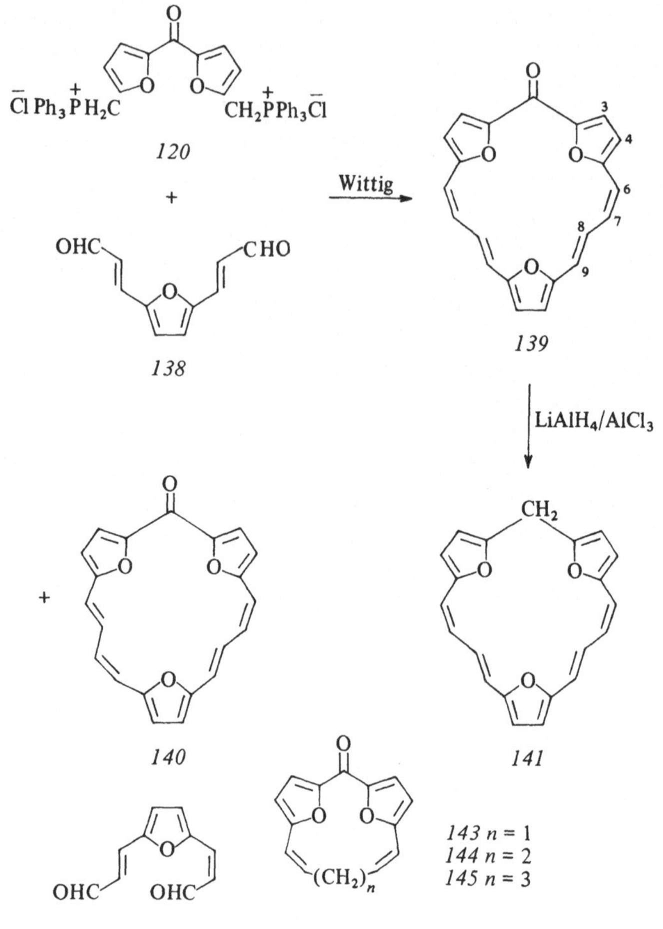

Oxidative coupling[55] of the acetylenic alcohol *146* under Eglinton conditions followed by acidic treatment of the product gave the aldehydes *147* (37%) and *148* (8%). Wittig reaction of aldehyde *147* and the bisphosphonium salt *120* with 1,5-diazabicyclo[4.3.0]non-5-ene as base gave the [21]annulenone *149*. This on reduction with lithium aluminium hydride-aluminium chloride gave the homoannulene

150. Comparisons between the NMR spectra of *149* and *150*, and the ketones *143*, *144*, and *145* indicated that the annulenone *140* is atropic.

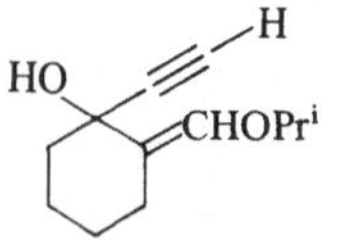

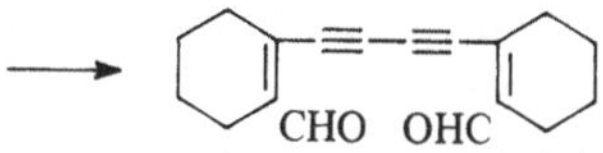

146

147 +

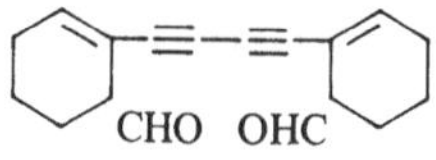

148

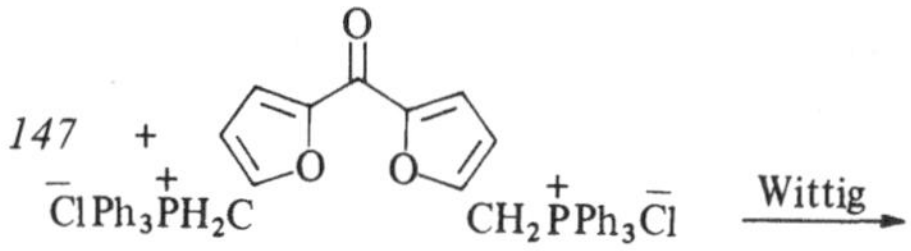

147 +

120

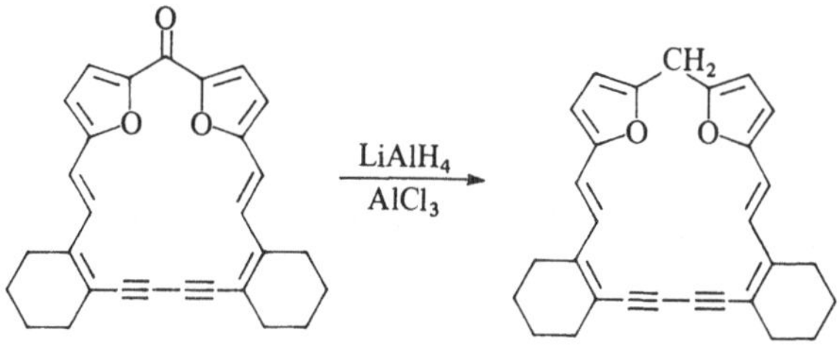

Wittig

149

LiAlH₄
AlCl₃

150

Annulenediones

Recent investigations have been concerned with determination of the extent of the quininoid nature of carbocyclic conjugated diketones.

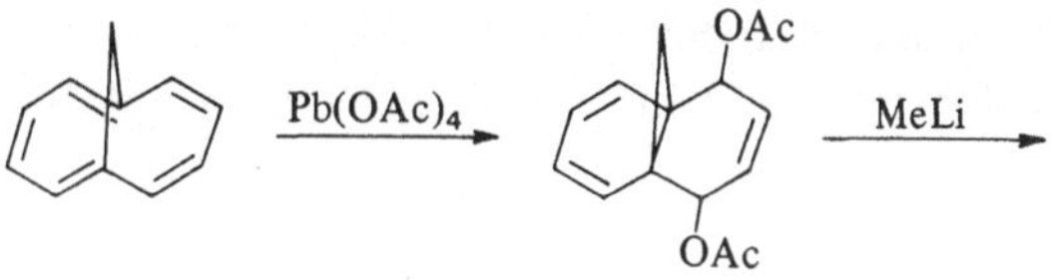

151

Pb(OAc)₄

152

MeLi

153

MnO₂

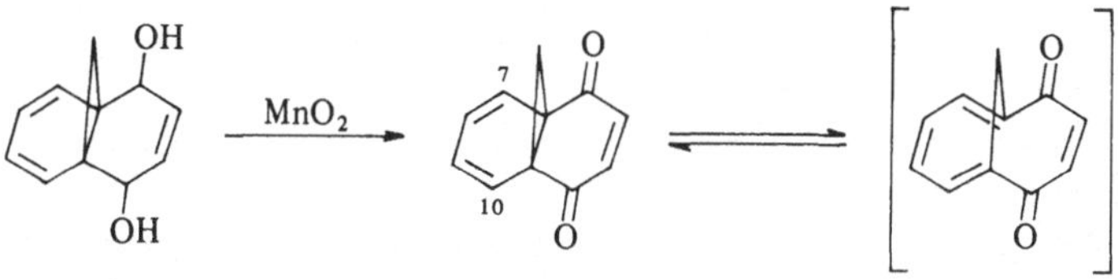

154

155

135

Lead tetraacetate oxidation of 1,6-methano[10]annulene *151* gave the *cis*-di-acetate *152* which from spectral evidence exists in the norcaradiene form[56]. The stereochemical relationship of the acetoxy groups is unknown. Treatment of *152* with methyllithium gave the diol *153* which on manganese dioxide oxidation gave the diketone *154* in quantitative yield. The coupling constants of H_7 and H_{10} and the geminal coupling constant of the methylene protons indicated that this di-ketone existed in the norcaradiene form *154* rather than the cycloheptatriene form *155*. Cyclopropane ring strain is considerably increased by geminal fluorine sub-stituents[57] and therefore replacement of the methylene protons of *154* with fluorine atoms should favour the cycloheptatriene structure *159* over the norcaradiene struc-

ture *160*. In the event oxidation of *156* with lead tetraacetate gave the *cis*-diacetate *157* which on treatment with methyl-lithium gave the diol *158*. In contrast to the previous case both *157* and *158* exist as the cycloheptatriene valence tautomers. Manganese dioxide oxidation of *158* gave the diketone *159* which from its ^{1}H and ^{19}F NMR spectra has the cycloheptatriene structure. Mild reduction of the tricyclic diketone *154* yielded the bicyclic semiquinone *161* the hyperfine splitting in the ESR

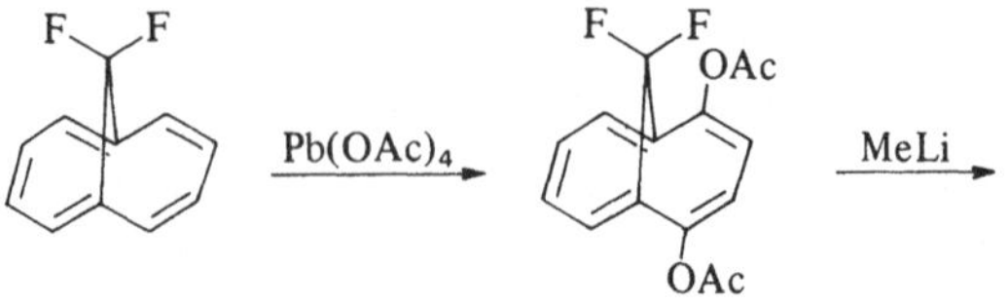

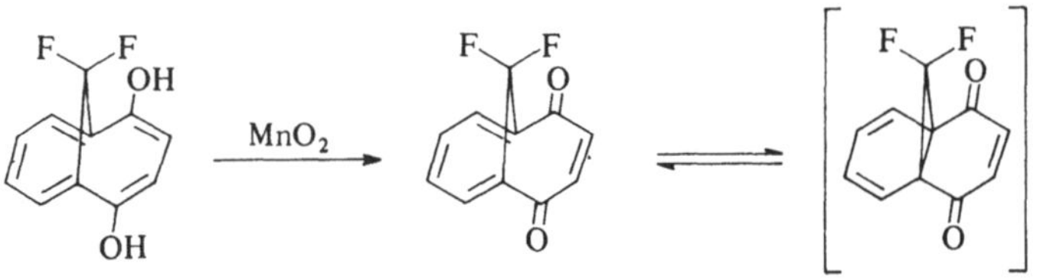

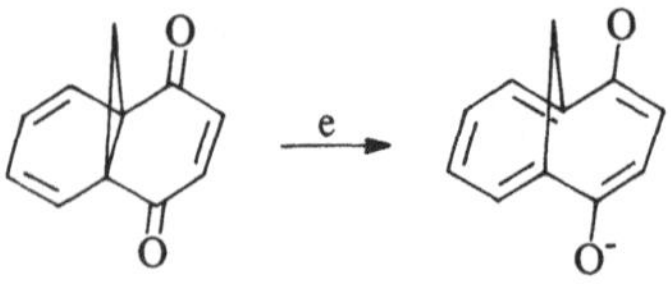

spectrum of which indicates that the unpaired electron is extensively delocalised into the π-electron system[58].

In contrast to *154* the isomeric bicyclo[4.4.1]undeca-3,5,8,10-tetraene-2,7-dione *166*[59] can only have a bicyclic structure. Formation of the bis-Grignard reagent *163* of the dibromo[10]annulene *162* was accomplished by reaction of the di-lithio derivative with anhydrous magnesium bromide. Treatment of *163* with tert-butyl perbenzoate gave the di-tert-butyl ether *164* which was cleaved with toluene-*p*-sulphonic acid in benzene to the diketone *165*. Although 2,3-dichloro-5,6-dicyano-*p*-benzoquinone did not dehydrogenate *165* to *166* the latter was obtained by allylic bromination of *165* with *N*-bromosuccinimide followed by debromination with potassium iodide in

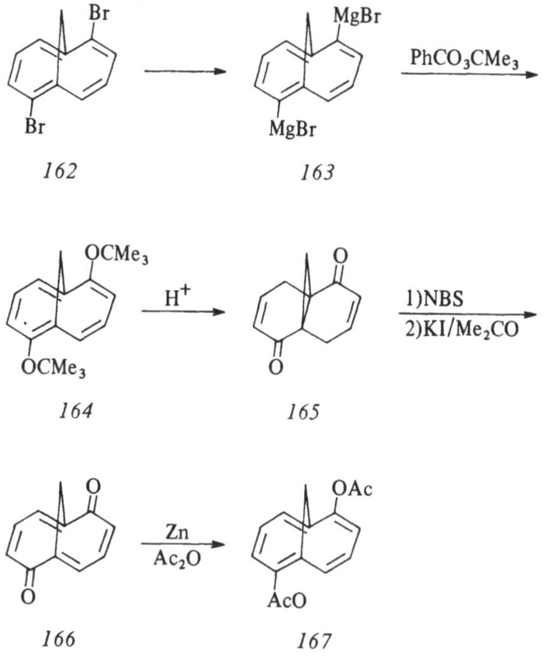

acetone. The diketone *166* behaved chemically as a quinone since it underwent reductive acetylation with zinc and acetic anhydride and afforded the diacetate *167*. The question as to whether the dione *166* is truly quininoid has not yet been resolved.

A number of macrocyclic diones related to tetradehydro[18]annulene have been synthesised. Eglington coupling of the α,β-unsaturated aldehyde *86* gave the dialdehyde *168* which on reaction with a large excess of ethynyl magnesium bromide afforded the diol *169*. Oxidation of this diol with manganese dioxide yielded the diketone *170* which on Glaser coupling have the cyclic dione *171*. In the NMR spectrum of *171* the internal protons H_3 resonate at τ 0.94 whereas in *170* the analogous protons resonate at 1.78. The downfield shift of the internal proton H_3 has been attributed to the greater deshielding in the rigid cyclic system *171* than in *170* by the surrounding olefinic and/or acetylenic bonds[60].

The alcohol *172* obtained by reaction of *86* with ethynyl magnesium bromide on oxidation with manganese dioxide gave the ketone *173*[61]. Glaser coupling of the ketone *173* gave an equimolar mixture of the two acyclic diketones *170* and *174*.

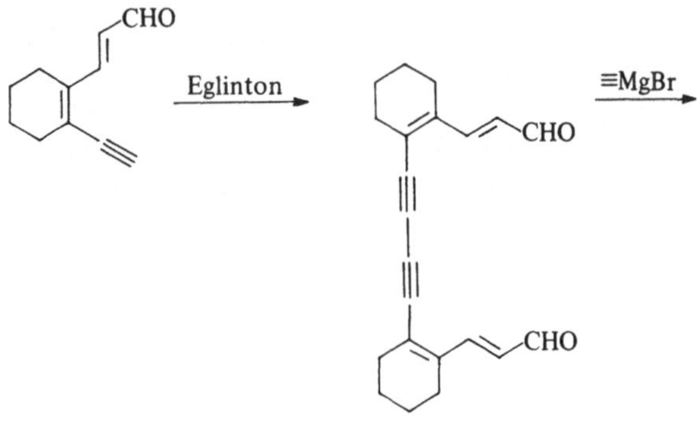

CHO

Eglinton

≡MgBr

CHO

CHO

86   *168*

OH

OH

MnO₂

O

O

Glaser

2

3

O

O

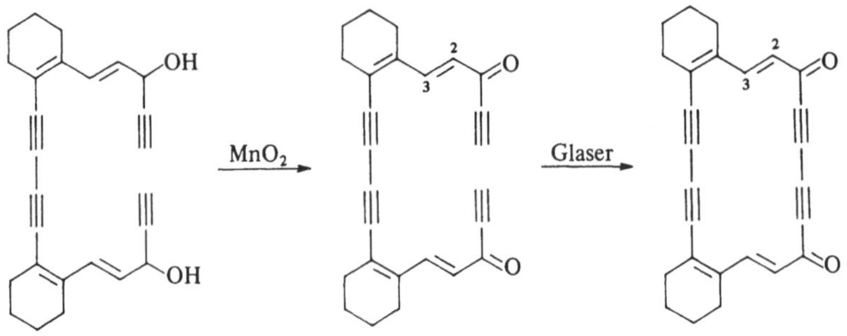

169   *170*   *171*

The dione *174* only underwent oxidative coupling under Hay's conditions[62] and afforded the cyclic dione *175*.

Cis-3-methyl-2-penten-4-ynal *72* was converted to the ketone *176* which on coupling under Glaser conditions led directly to the annulenediones *177* and *178*[61].

It is known that the reduction potentials of quinones are related to the aromatic stabilization of the parent conjugated systems. In an attempt to relate the annulenediones *171*, *177*, and *178* to the tetradehydro[18]annulene system Breslow and co-workers[63] have studied their electrochemical reduction by cyclic voltammetry. These diones can easily be reduced to the corresponding dianions, e.g. 171 → 179. These

Table 1. Electrochemical reduction potentials

Compound	E_1	E_2	$E_1 - E_2$	$E_1 + E_2$
p-Benzoquinone	−0.52	−1.40	0.88	−1.92
1,4-Napthoquinone	−0.64	−1.47	0.83	−2.11
9,10-Anthraquinone	−0.93	−1.63	0.70	−2.56
171	−0.70	−1.04	0.34	−1.74
177	−0.68	−0.94	0.25	−1.61
178	−0.66	−0.93	0.28	−1.60

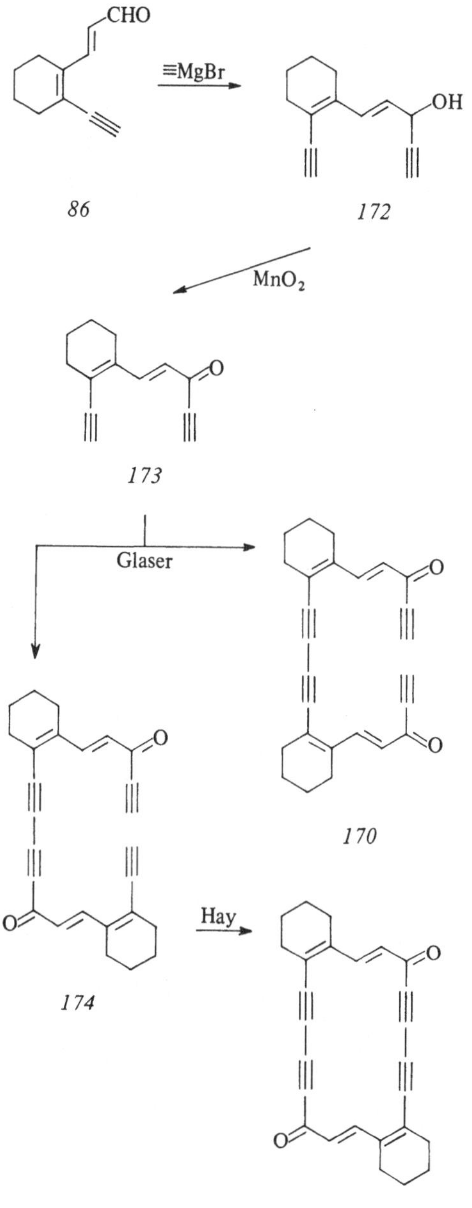

CHO

$\equiv$MgBr

OH

86

172

MnO$_2$

173

Glaser

170

174

Hay

175

compounds showed chemical and electrochemical reversibility at both the first and second waves which correspond to the radical anion and dianion respectively. The sum of the potentials for the first and second waves $(E_1 + E_2)$ is related to the total energy change involved in the conversion of the dione to the dianion. Reduction of p-benzoquinone is more easy than reduction of 1,4-napthoquinone which is in turn easier than reduction of 9,10-anthraquinone thus reflecting the increase in aromatic stabilization on reduction of the quinone to the respective aromatic hydroquinone

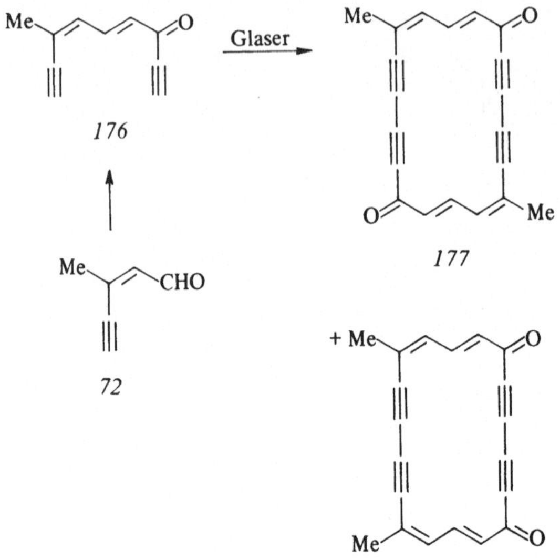

176 177 72 178

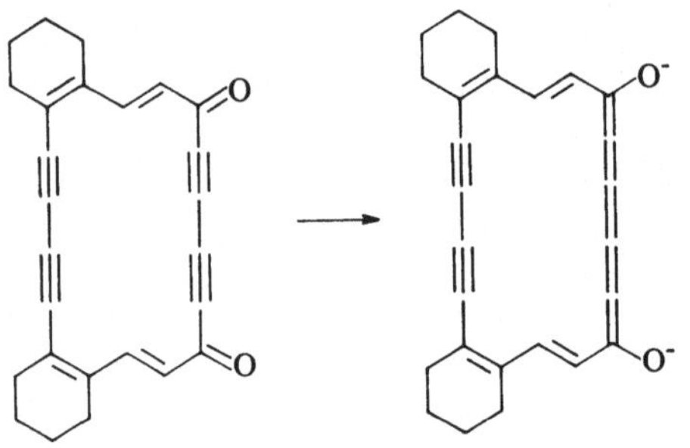

171 179

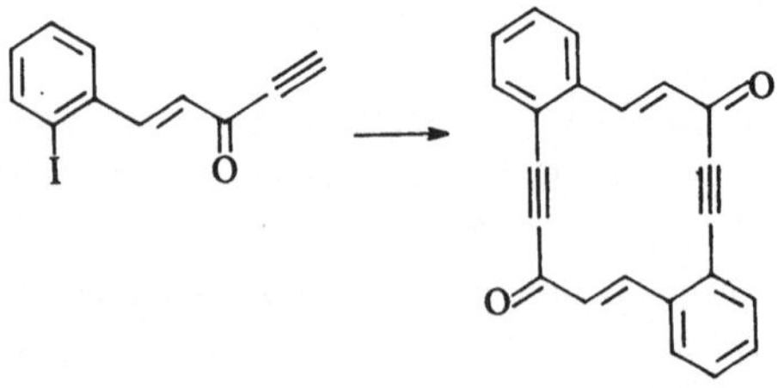

180 181

dianion. Comparison of the $E_1 + E_2$ values for the diones *171*, *177*, and *178* with that for *p*-benzoquinone would indicate that the tetrahydro[18]annulene systems derived from *171*, *177*, and *178* are considerably more stabilised (0.3 eV, 7 kcal mole^{-1}) than benzene by cyclic π-electron delocalisation. The potentials are also influenced by electrostatic repulsion effects as is seen for the difference in reduction potential for the first and second waves ($E_1 - E_2$). This difference is much larger for smaller molecules and frustrates attempts to relate $\Delta(E_1 + E_2)$ with resonance energy. It was argued that the almost identical potentials for *177* and *178* were due to the carbonyl groups in these systems being too far apart for much electrostatic inter-action in either case. That *171*, *177*, and *178* are easily reduced indicates that they are indeed quinones of aromatic systems.

The benzo[14]annulenedione *181* has been prepared by Castro reaction of the cuprous salt of the *o*-iodocinnamoylacetylene *180*[64].

Conclusions

It is apparent that the higher annulenones exhibit diatropicity and paratropicity to a much smaller extent than the annulenes of a similar ring size, although the hydroxy-annulenium ions are comparable in this respect with the annulenes. The higher an-nulenones are also sensitive to perturbations, quite small alterations in structure are sufficient to destroy the diatropicity and paratropicity of these systems. This is in keeping with the diminished magnitude of these effects as compared with the an-nulenes. The dichotomy in ground state properties between the bridged [11]an-nulenones, which are atropic, and the higher annulenones may find an explanation in this enhanced susceptibility to perturbation.

No adequate theoretical treatment of the higher annulenones which explains their NMR properties has yet emerged. It is to be hoped that this situation will be remedied in the near future. As yet little work on the X-ray crystal structures of the higher annulenones has appeared and this would also be a profitable field of investiga-tion[65]. The question as to whether it is possible to synthesise a higher annulenone which does not possess such structural perturbations as bridging groups or acetylenic bonds is still open, and offers considerable challenge. Tropone is still the largest parent annulenone known. The question of the limiting ring size for diatropicity and paratropicity in annulenones also awaits resolution.

The field of annulenediones will probably continue to prove an intriguing area of research.

Acknowledgements. We thank Professors F. Sondheimer and E. Vogel for the communication of unpublished results. We are indebted to Professor F. Sondheimer and Dr. P. J. Garratt for reading the article in manuscript and for their many valuable comments.

We thank the Australian Research Grants Committee for financial support.

141

References

[1] a) Sondheimer, F.: Accts. Chem. Res. *5*, 81 (1972);
   b) Sondheimer, F.: Proc. Roy. Soc. (London). *297A*, 173 (1967);
   c) Sondheimer, F. *et al.*,: Chem. Soc. Spec. Publ. *21*, 75 (*1967*);
   d) Sondheimer, F.: Pure Appl. Chem. *7*, 363 (1963).

[2] Boekelheide, V., Pepperdine, W.: J. Am. Chem. Soc. *92*, 3684 (1970) and earlier papers.

[3] a) Vogel, E.: Chem. Soc. Spec. Publ. *21*, 113 (1967);
   b) Vogel, E.: Chimia *22*, 21 (1968);
   c) Vogel, E.: Proc. Robert A. Welch Found. Conf. Chem. Res. *12*, 215 (1968).

[4] Dauben, Jr., H. J., Wilson, J. D., Laity, J. L., in: Non-benzenoid aromatics (ed. J. P. Snyder), Vol 2, p. 167. New York: Academic Press 1971.

[5] Musher, J. I.: J. Chem. Phys. *43*, 4081 (1965).

[6] Nowakowski, J.: Theor. Chim. Acta *10*, 79 (1968).

[7] Garratt, P. J., Rowland, N. E., Sondheimer, F.: Tetrahedron *27*, 3157 (1971).

[8] Waugh, J. S., Fessenden, D. W.: J. Am. Chem. Soc. *79*, 846 (1957).

[9] Longuett-Higgins, H. C., Salem, L.: Proc. Roy. Soc. (London) *251A*, 172 (1959).

[10] Dewar, M. J. S., Gleicher, G. J. Amer. Chem. Soc. *87*, 685 (1965).

[11] Metcalf, B. W., Sondheimer, F.: J. Am. Chem. Soc. *93*, 5271 (1971).

[12] Iyoda, M., Nakagawa, M.: Tetrahedron Letters. *1973*, 4743.

[13] Breslow, R.: Accts. Chem. Res. *6*, 393 (1973).

[14] Dewar, M. J. S.: Chem. Soc. Spec. Publ. *21*, 177 (1967).

[15] Longuett-Higgins, H. C.: Chem. Soc. Spec. Publ. *21*, 109 (1967).

[16] Pople, J. A., Untch, K. G.: J. Am. Chem. Soc. *88*, 4811 (1966).

[17] Baer, F., Kuhn, H., Regel, W.: Z. Naturforsch. *22A*, 103 (1967).

[18] Pilling, G. M., Sondheimer, F.: J. Am. Chem. Soc. *90*, 5610 (1968); *93*, 1977 (1971).

[19] Hess Jr., B. A., Schaad, L. J., Holyoke, C. W., Jr.: Tetrahedron *28*, 5299 (1972).

[20] Breslow, R, Eicher, T., Krebs, A., Peterson, R. A., Posner, J.: J. Am. Chem. Soc. *87*, 1320 (1965).

[21] Breslow, R., Altmann, L. J., Krebs, A., Mechacsi, E., Murata, I., Peterson, R. A., Posner, J.: J. Am. Chem. Soc. *87*, 1326 (1965).

[22] Breslow, R., Ryan, G.: J. Am. Chem. Soc. *89*, 3073 (1967).

[23] Krebs, A.: Angew. Chem. *77*, 10 (1965); Angew. Chem. Intern. Ed. *4*, 10 (1965).

[24] Tobey, S. B., in: Aromaticity, pseudoaromaticity, antiaromaticity (ed. E. D. Bergmann and B. Pullman), p. 351. Jerusalem: Israel Academy of Sciences and Humanities 1971.

[25] Chapman, O. L., McIntosh, C. L.: Chem. Commun. *1971*, 770.

[26] Garbisch Jr., E. W., Sprecher, R. F.: J. Am. Chem. Soc. *91*, 6785 (1969).

[27] Dauben Jr., H. J., Ringold, H. J.: J. Am. Chem. Soc. *73*, 876 (1951); Doering, W. von E., Detert, F. L.: J. Am. Chem. Soc. *73*, 876 (1951).

[28] Kimura, K., Sazuki, S., Kimura, M., Kubo, M.: J. Chem. Phys. *27*, 320 (1957); Bull. Chem. Soc. Japan *31*, 1051 (1958).

[29] Nozoe, T., Mukai, T., Takase, K., Nagase, T.: Proc. Japan Acad. *28*, 477 (1972).

[30] Giacomo, A. D., Smyth, C. P.: J. Am. Chem. Soc. *74*, 4411 (1952); Kurita, Y., Seto, S., Nozoe, T., Kubo, M.: Bull. Chem. Soc. Japan *26*, 272 (1953).

[31] Bertelli, D. J., Andrews, T. G., Jr.: J. Am. Chem. Soc. *91*, 5280 (1969); Bertelli, D. J., Andrews, T. G., Crews, P. O.: J. Am. Chem. Soc. *91*, 5286 (1969).

[32] Dewar, M. J. S., Harget, A. J., Trinajastić, N.: J. Am. Chem. Soc. *92*, 6335 (1970); Dewar, M. J. S., Trinajastić, N.: Croatica Chemica Acta, *42*, 1 (1970).

[33] Barrow, M. J., Mills, O. S.: J. C. S. Chem. Commun. *1973*, 66.

[34] Grimme, W., Reisdorff, J., Jünemann, W., Vogel, E.: J. Am. Chem. Soc. *92*, 6335 (1970).

[35] Vogel, E.: Personal communication.

[36] Hudson, D. W., Mills, O. S.: Chem. Commun. *1971*, 153.

[37] Ogawa, H., Kato, H., Yoshida, M.: Tetrahedron Letters *1971*, 1793.

[38] Pilling, G. M., Sondheimer, F.: J. Am. Chem. Soc. *93*, 1971 (1970).

39) Howes, P. D., LeGoff, E., Sondheimer, F.: Tetrahedron Letters *1972*, 3691.
40) Staab, H. A., Bader, R.: Chem. Ber. *103*, 1157 (1970).
41) Sondheimer, F.: Private communication.
42) Nakagawa, M., Iyoda, M., Kasiwadani, T.: 30th Meeting of the Chemical Society of Japan: Osaka: Spring 1974.
43) Cotterrell, G. P., Mitchell, G. H., Sondheimer, F., Pilling, G. M.: J. Am. Chem. Soc. *93*, 259 (1971).
44) Howes, P. D., LeGoff, E., Sondheimer, F.: Tetrahedron Letters *1972*, 3695.
45) Ogawa, H., Shimojo, N., Yoshida, M.: Tetrahedron Letters *1971*, 2013.
46) Ogawa, H., Yoshida, M., Saikachi, H.: Tetrahedron Letters *1972*, 153.
47) Ogawa, H., Ibii, N., Kato, H., Tabushi, I., Cresp, T. M., Sargent, M. V.: unpublished work.
48) Brown, G. W., Sondheimer, F.: J. Am. Chem. Soc. *91*, 670 (1969).
49) Calder, I. C., Gaoni, Y., Sondheimer, F.: J. Am. Chem. Soc. *90*, 4946 (1968).
50) Griffiths, J., Sondheimer, F.: J. Am. Chem. Soc. *91*, 7518 (1969).
51) Sondheimer, F.: Pure Appl. Chem. *28*, 331 (1971).
52) Cresp, T. M., Sargent, M. V.: J. C. S. Perkin 1 *1973*, 2961.
53) Vogel, E., Pretzer, W., Böll, W. A.: Tetrahedron Letters *1965* 3613.
54) Abraham, R. J., Jackson, A. H., Kenner, G. W.: J. Chem. Soc. *1961*, 3468; Woodward, R. B., Skaric, V.: J. Am. Chem. Soc. *83*, 4676 (1961).
55) Cresp, T. M., Sargent, M. V.: J. C. S. Chem. Commun. *1974*, 101; J. C. S. Perkin 1: in press.
56) Vogel, E., Lohmar, E., Böll, W. A., Söhngen, B., Müller, K., Günther, H.: Angew. Chem. *83*, 401 (1971); Angew. Chem. Intern. Ed. *10*, 398 (1971).
57) Mitsch, R. A., Neuvar, E. W.: J. Phys. Chem. *70*, 546 (1966); Günther, H.: Tetrahedron Letters *1970*, 5173.
58) Russell, G. A., Ku, T., Lokensgard, J.: J. Am. Chem. Soc. *92*, 3833 (1970).
59) Vogel, E., Böll, W. A., Lohmar, E.: Angew. Chem. *83*, 403 (1971); Angew. Chem. Intern. Ed. *10*, 399 (1971).
60) Yamamoto, K., Sondheimer, F.: Angew. Chem. *85*, 41 (1973); Angew. Chem. Intern. Ed. *12*, 68 (1973).
61) Darby, I. N., Yamomoto, K., Sondheimer, F.: J. Am. Chem. Soc. *96*, 248 (1974).
62) Hay, A. S.: J. Org. Chem. *27*, 3320 (1962).
63) Breslow, R., Murayama, D., Drury, R., Sondheimer, F.: J. Am. Chem. Soc. *96*, 249 (1974).
64) Kojima, T., Sakata, T., Misumi, S.: Bull. Chem. Soc. Japan *45*, 2834 (1972).
65) See however: Duckworth, V. F., Hitchcock, P. B., Mason, R.: Chem. Commun. *1971*, 963.

Received September 17, 1974

Springer-Verlag
Berlin Heidelberg New York
München Johannesburg London
Madrid New Delhi
Paris Rio de Janeiro Sydney Tokyo Utrecht Wien

POLYMER CHEMISTRY

By Professor B. Vollmert,
Polymer Institute, University of Karlsruhe, Germany
Translated from the German by E. H. Immergut, New York
With 630 figures. XVII, 652 pages. 1973
Cloth DM 72.—; US $29.40 ISBN 3-540-05631-9
Prices are subject to change without notice

This book gives a comprehensive coverage of the synthesis of polymers and their reactions, structure, and properties. The treatment of the reactions used in the preparation of macromolecules and in their transformation into cross-linked materials is particularly detailed and complete. The book also gives an up-to-date presentation of other important topics, such as enzymatic and protein synthesis, solution properties of macromolecules, polymer in the solid state. The content and presentation of Professor Vollmert's book is more encompassing than most existing treatises, and its numerous figures and tables convey a wealth of data, never, however, at the expense of intellectual clarity or educational value.
The presentation is mainly on a fundamental and general level and yet the reader—student or professional—is gradually and almost casually introduced to all important natural and synthetic polymers. Complicated phenomena are explained with the aid of the simplest available examples and models in order to ensure complete understanding. However, the reader is also encouraged to think for himself and even to criticize the author's point of view. All of the chapters have been revised and enlarged from the German edition, and many of the sections are entirely new.

Contents: Introduction. — Structural Principles. — Synthesis and Reactions of Macromolecular Compounds. — The Properties of the Individual Macromolecule. — States of Macromolecular Aggregation.

A. Gossauer **Die Chemie der Pyrrole**

17 Abbildungen. XX, 433 Seiten. 1974
(Organische Chemie in Einzeldarstellungen, Band 15)
Gebunden DM 158,–; US $68.00
ISBN 3-540-06603-9
Preisänderungen vorbehalten

Inhaltsübersicht: Struktur des Pyrrol-Moleküls. – Analytische Methoden. – Reaktivität der Pyrrole. – Pyrrol-Metall-Derivate. – Pyrrole als Naturprodukte. – Pyrrol-Ringsynthesen. – Synthetische Methoden.

Das Pyrrol und seine Derivate haben als technische Grundstoffe wie auch als Naturprodukte wachsendes Interesse gewonnen. – Diese Monographie ist eine umfassende Übersicht über die seit 1934 erschienene Literatur (ausgenommen Porphyrine). Bedingt durch die seitdem ständig wachsende Anzahl der Veröffentlichungen, die sich mit den physikalischen Eigenschaften dieser Verbindungsklasse befassen, weichen Konzeption und Gliederung dieses Buches von denjenigen des klassischen Werkes von H. Fischer und H. Orth grundsätzlich ab. Die Anwendung quantenmechanischer Rechenverfahren zur Deutung der Eigenschaften des Pyrrol-Moleküls wird im ersten Kapitel ausführlich erörtert. Die entscheidende Bedeutung der physikalischen Methoden zur Untersuchung der Konstitution und Reaktivität des Pyrrols und seiner Derivate ist durch zahlreiche tabellarisch geordnete Datenangaben, deren Interpretation im Text diskutiert wird, hervorgehoben. Dem präparativ arbeitenden Chemiker soll die Systematisierung der synthetischen Methoden bei der Suche nach der einschlägigen Literatur helfen: Ringsynthesen sind nach dem Aufbaumodus des Heterocyclus, die Einführung von Substituenten nach funktionellen Gruppen klassifiziert und anhand von Schemata übersichtlich zusammengefaßt worden. Bei der Zusammenstellung der Abbildungen wurden neben den trivialen Beispielen, die zum besseren Verständnis des Textes dienen, besonders jene Reaktionen ausgewählt, bei denen Pyrrole Ausgangsverbindungen zur Darstellung anderer Heterocyclen (Indole, Pyrrolizine, Azepine, u.a.) sind. Besondere Sorgfalt gilt der Beschreibung von Reaktionsmechanismen. (2621 Literaturzitate.)

Springer-Verlag
Berlin
Heidelberg
New York